LOST WORLD

LOST WORLD

REWRITING PREHISTORY— HOW NEW SCIENCE IS TRACING AMERICA'S ICE AGE MARINERS

TOM KOPPEL

ATRIA BOOKS
NEW YORK LONDON TORONTO SYDNEY SINGAPORE

1230 Avenue of the Americas
New York, NY 10020

For information address Atria Books, 1230 Avenue
of the Americas, New York, NY 10020

ISBN: 0-7434-5357-3

First Atria Books hardcover printing June 2003

10 9 8 7 6 5 4 3 2 1

Printed in the U.S.A.

For information regarding special discounts for bulk purchases,
please contact Simon & Schuster Special Sales at 1-800-456-6798
or business@simonandschuster.com

Dedicated to Doris Phillips,
my "second mum," whose love and support
have meant so much to me over so many years.

ACKNOWLEDGMENTS

THIS BOOK EMERGED from intermittent periods of work, starting in 1985, in which I followed the evolving debate over the origins of the First Americans and the related search for evidence of ancient people on the North Pacific coast. The effort took me on a series of research trips and resulted in a number of magazine and newspaper articles. I also interviewed many scientists by phone and received extensive published materials from others. The project has stretched over such a long period that not all may even remember me, but I remember them and remain grateful for their help.

I am especially indebted to the scientists who invited me to join them at the three most significant coastal research sites and gave me so much of their time and attention. They are: Heiner Josenhans, Daryl Fedje, Jim Dixon, Tim Heaton, and Don Morris.

Others who granted me interviews (in person, by phone, or by e-mail), sent me documentary material, or simply helped me make other important contacts, include: Terry Fifield, Vaughn Barrie, Martin Magne, Quentin Mackie, Kim Conway, Ivan Frydecky, Peter Waddell, Pete Smith, Tina Christensen, Ian Sumpter, C. Loring Brace, Henk Don, John Johnson, Jim Ringer, Norm Easton, Ruth Gruhn, Grant Keddie, Christy Turner II, Lionel Jackson, Emöke Szathmary, Richard Morlan, Jacques Cinq-Mars, John Clague, Richard Hebda, Alan Thorne, Knut Fladmark, Geoffrey Irwin, Rolf Mathewes, Fumiko Ikawa-Smith, Joanne McSporran, Tom Greene Jr., Ernie Gladstone, Tucker Brown, Captain Gold, Jordan Yeltatzie, Sean Young, Barbara Wilson, Keith Rowsell, Barb Rowsell, Tom Ager, Brent Ward, Kei Nozaki, Joe Neynaber, Bob Gray, Wade Sirles, Janna Carpenter, Craig

Lee, Dylan Reitmeyer, Heather Mrzlack, Heidi Manger, Patrick Olsen, Eric Parrish, Simon Easteal, Peter Bellwood, and Mel Aikens.

I also owe thanks to fellow writers and photographers who offered advice or helped in a variety of other ways, as well as to former agents who shaped my thinking about the book project, and to the editors who assigned and published the articles I wrote along the way. They include: Brian Fagan, Heather Pringle, Sid Tafler, Ken Garrett, Mike Parfit, Suzanne Chisholm, Sandy MacDonald, Jennifer Barclay, Jan Whitford, Stewart Muir, Neil Reynolds, Ian Darragh, Eric Harris, Rick Boychuk, Alan Morantz, and Michael Bawaya.

I am very grateful to the British Columbia Arts Council for the generous grant that allowed me to get started on the book itself.

I owe a particular debt to my agent Dorian Karchmar and my editor Brenda Copeland for their high literary standards, gentle prodding, thoughtful critiques, and overall attention to detail.

And finally, heartfelt thanks to my wife, Annie, for the time she spent proofreading the many drafts, assisting in countless other ways, and coping with my restless author's nights.

CONTENTS

Prologue xi
Chapter One: *Cave of the Bears* 1
Chapter Two: *Northern Connection* 12
Chapter Three: *Important Artifacts Found* 26
Chapter Four: *Clovis First* 43
Chapter Five: *Coastal Network* 61
Chapter Six: *Barnacles and Bones* 77
Chapter Seven: *Orcas Bring Good Luck* 98
Chapter Eight: *Ancient Mariners* 120
Chapter Nine: *Archaeology Gets Its Feet Wet* 134
Chapter Ten: *Long Chronology* 147
Chapter Eleven: *Maverick Archaeologist* 164
Chapter Twelve: *Boulders That Talk* 181
Chapter Thirteen: *Needle in a Haystack* 192
Chapter Fourteen: *Archaeology's Gold Standard* 204
Chapter Fifteen: *All Alone Stone* 221
Chapter Sixteen: *Arlington Woman* 241
Chapter Seventeen: *Emerging Consensus* 258
Chapter Eighteen: *Ancient Odyssey* 279
Index 289

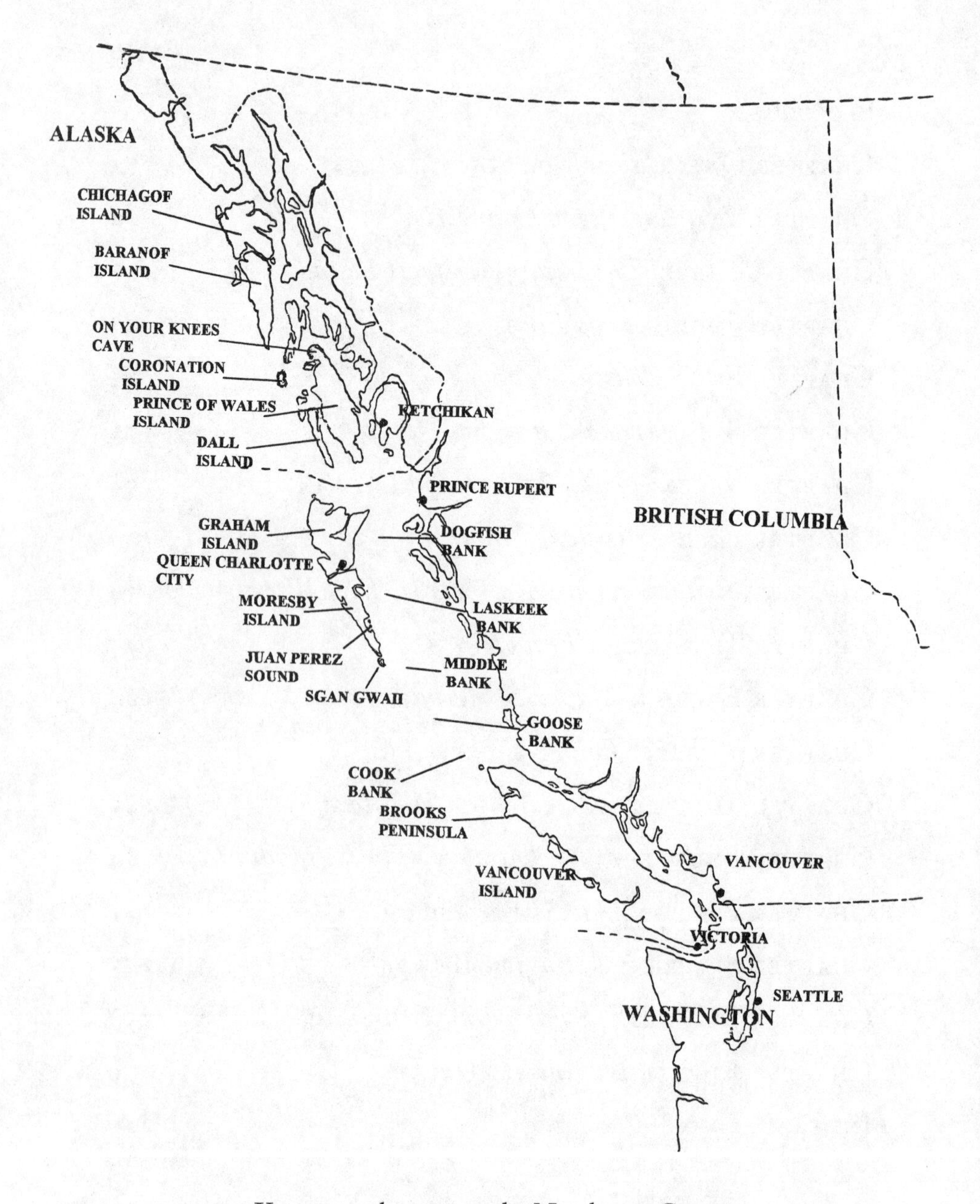

Key research areas on the Northwest Coast.

PROLOGUE

A Drowned Forest

THE NORTHWEST COAST exudes more than a whiff of mystery and adventure. Brooding forests line the shores of fjords that slash deep into the mountain fastness. Haunted faces on carved cedar poles preside in silent disdain over abandoned native villages. The hushed secrets of the past beckon, yet remain elusive. Most enigmatic of all, lying hidden beyond the outer beaches and reefs and far below the surface is an entire lost world, a domain of the deep that is evoked in ancient native myth and confirmed by the latest technological wizardry. Invisible to modern eyes, this drowned zone holds clues to ancient seafaring people and how they came to populate our hemisphere.

These forgotten coastal dwellers first came to life for me when I rented a seaside homestead on a small island in British Columbia. The Stone House, as the owner called it, stood in a clearing just behind the beach at the head of an isolated bay fringed with regal Douglas firs and twisted red madrone trees. I liked the house, but what really intrigued me was the spot it stood on, a grassy level platform that spanned a quarter acre and was raised about six feet above the high tide line. As I paced around it with my aged landlord, he told me that this platform was no accident of geology but a human creation. It was an Indian shell midden, or organic native garbage dump, a dense buildup of shells that resulted from thousands of years of seasonal shellfish harvesting. There were similar middens behind nearly every beach along the sheltered "Inside Passage," which runs from Puget Sound in Washington State right up into the panhandle of Alaska.

My landlord had unearthed a crumbling human skull when he dug the house foundations four decades earlier. During my years there I was humbled by the thought of living on native ground hallowed by the centuries. Soon, I planted a vegetable patch on the midden next to the house. Digging in, I discovered that the soil, far from being the mixed sand and clay of ordinary garden dirt, consisted of crushed and broken shell along with humus from fallen leaves and droppings from the island's deer, sheep, and goats. My produce thrived.

One day, I hiked across the island to another bay and came upon a scene that helped me understand how middens were formed. A family from one of the Coast Salish tribes was busy digging clams by the bushel-load at low tide to sell to cash buyers. Their boat was pulled up high and dry on the mud flats. Lined up off to one side were a dozen or more thigh-high gunny sacks full of clams, each almost too heavy to lift. I was amazed at the volume of shellfish that could be taken from a single small beach, and it was easy to see how over scores of centuries shell middens could build up to depths of five to ten feet or more. In these modern times, the succulent bivalves would be taken away and shucked at a processing plant, or sold live in the shell. In the past, though, the Indians simply camped out at spots like this for days or weeks at a time, and were still doing so when the first European settlers arrived. After digging the clams, they smoked or roasted them and skewered them on sticks for storage. Shellfish gathering was a major part of their seasonal harvest cycle and was essential for laying in a supply of food for the winter. The shells were simply left behind above the tide line.

I had majored in anthropology briefly in college before moving on to other things, and living on the midden rekindled a passionate interest. It wasn't long before I was wangling invitations to potlatch ceremonies, going to watch native fisheries, and generally delving into coastal anthropology. Poking around in academic journals published by universities in Seattle and Vancouver, I learned that the leaching of chemicals from the shells helped greatly to preserve the bone and antler artifacts, such as fine harpoon points, that were so important to prehistoric coastal hunting and gathering. Nearly all coastal archaeology, in fact, involved digging in shell middens. This led me to other middens and archaeological sites on the nearby islands.

I quickly discovered that looking at today's shores could only take me back about 5,000 years, at least in my area. Because sea level on the Northwest Coast has changed so drastically and so rapidly since the Ice Age, in most places anything older was under water. Coastal archaeology along the Inside Passage had barely scratched the surface of land and sea, but the implications of ancient sea level change went well beyond creating local difficulties for a handful of anthropology professors.

At the peak of the last glaciation of the Ice Age—18,000 to 20,000 years ago—great ice sheets covered much of northern Europe and Russia. In the southern hemisphere, enormous glaciers mantled the Andes. And in North America, the Laurentide ice sheet, with a central dome as much as three miles thick, extended from eastern Canada almost to the Rockies and down into the United States beyond the Great Lakes. Another ice sheet, the Cordilleran, stretched over the Rockies to the coast of Alaska, British Columbia, and northern Washington State. So much water was locked up in ice that world sea level was 350 to 400 feet lower than it is today.

Entire bodies of water that had existed prior to the glaciation dried up, and the former sea bottom became exposed land. In the best known case, the Bering Strait disappeared and became a broad "land bridge," around a thousand miles wide, connecting Asia and North America. We are all familiar with schoolbook pictures of ancient people, dressed in furs, hunting mammoth elephants and trekking across the land bridge to populate the New World.

There were also many islands, in all the world's oceans, that no longer exist. On maps of the North Pacific, we find no islands between Hawaii and the mainland of Alaska and British Columbia except just off the mainland coast. But in the last glaciation there were several islands well out in the Pacific. These were the tops of submarine mountains coming up from very deep water, just like the volcanic islands of Hawaii. Now they lie a hundred or more feet below the surface and attract vast schools of feeding fish. Oceanographers call them seamounts. Other islands that still exist at today's sea level were simply much larger at that time, and in some cases were connected to the mainland.

Nearly all the world's coastlines lay much farther to seaward than they do now. One respected archaeologist has calculated that the total amount of "extra" land this lowering of sea level created along the coasts of North America was equal to one and one half times the area of Texas. Life on the exposed land was very different from what we can see in modern times. The Northwest Coast today has a mild, wet climate and dense coastal rain forest. At the peak of glaciation, though, there was pack ice along shore and tundralike vegetation on land, much like the present north coast of Alaska.

Then the Ice Age waned. The ice sheets melted and gradually retreated from the coast, and the sea rose worldwide. In some places there were odd local effects. Where shorelines had been pressed down by the weight of the ice, the land rebounded at times even faster than the rise of the world ocean. But overall, between around 17,000 and 10,000 years ago, the rising sea simply drowned these once-extensive coastal lands under hundreds of feet of water.

Any people living along the coast would have been forced to relocate. But of course, there were no cities, probably not even permanent settlements of any size. People lived by hunting, fishing, and gathering, and they were always on the move, in step with their sources of food. And since it took thousands of years for the sea to rise, there was likely little disruption to the lives of those ancient people.

But there *is* one lingering consequence for us today. The rising sea inundated nearly all traces of Ice Age coastal dwellers throughout the world, making it nearly impossible to find evidence of when, where, and how people lived on those drowned lands. Because underwater searches are difficult, incredibly expensive and even, at times, dangerous, until recently archaeologists have hardly even tried. Instead, they have focused on what they *could* find, which is invariably at sites well inland from where so many people presumably lived. In this way, the rising sea has severely skewed the archaeological record and given a strong terrestrial bias to our image of the people and lifeways of these late glacial times.

We know, for example, that some late Ice Age people hunted mammoths with spears, because the stone spear points have been found in close association with mammoth bones. But they have been found on terra firma, of course, and far inland from the former ancient shoreline. We hardly know anything about what people were doing on the coasts of the world ten to twenty thousand years ago. Where did they live? What animals did they hunt? What fish did they catch, and how? What critters did they gather in the intertidal zone? What kinds of weapons and tools did they use?

For those of us who live in North America, there are additional questions. When did the first people arrive? From where and by what route? And for a few pioneering scientists of the last decade, that question has led to a more pressing one: Is it conceivable that the coastlines were inhabited *before* the interior of the continent? And if so, when?

Once I realized how drastically our coastal world had changed, everything I saw took on a slightly different hue. During those years I rebuilt a lovely old wooden cruising sailboat and spent much of each summer exploring the Northwest Coast with friends. While studying the undersea contours on nautical charts, I tried to picture the ancient world with hundreds of feet of water peeled away. The locations of beaches and sandspits, of islands and reefs—all would have been different. It must have been a rugged land of rubble-strewn moraines and tundra, where towering cliffs of ice fronted on the sea and roiling waves of ice fog drifted out over the continental shelf, a coastal fringe inhabited by polar bears, walruses, and other Arctic species that vanished from the region 10,000 years ago.

On one of my sailing trips north, we anchored for the evening among a maze of heavily wooded and unpopulated islands not far south of Alaska. We had seen only one other boat the entire day and easily imagined ourselves to be moored at the edge of the world. After dinner, my friend and I set out in the dinghy to explore our silent domain. At one island we went ashore and stumbled upon a well-concealed campsite littered with rusting pans and disintegrating plastic. A derelict canoe, cracked and hidden in the bush, was half filled with scummy water and mosquito larvae. Our fantasies ran wild. Was this the lair of a Crusoe-like recluse, or a fugitive from the law? Were there skeletons, literal or figurative, hidden on that island? It was a mystery we never solved.

A nearby island posed a different kind of riddle. Rowing along its shore, we noticed that the tide seemed to be exceptionally high. It was so high, in fact, that the sea was actually lapping at the base of dead trees on the gently sloping shoreline. Looking closer, we realized that all the trees from the high tide line back to at least 40 or 50 feet were gray and bare of foliage, killed off by the salt water. But the trees had not yet fallen over. We realized that this forest could not have become established in the first place if salt water had always soaked the soil in which they grew. The sea must have risen at least several feet within the lifespan of these very modestly sized trees, perhaps one hundred years.

Overall, of course, the Pacific Ocean itself was rising at no such rate, if at all. Rather, the land in this area was rapidly subsiding. The movement of tectonic plates was pinching and prodding and reconfiguring the entire Northwest Coast, dragging the land downward in some places, pushing it up in others. For me this was a graphic lesson that yanked regional sea level change out of the geological time scale of thousands or millions of years. Rising seas could change the location and character of the shoreline within a single lifetime, in what to geologists is the mere blink of an eye. A whole new element in coastal prehistory became real and relevant for me.

This was in the late 1980s, when just a few archaeologists were questioning the prevailing idea of how and when the First Americans arrived. Was it possible that the long-held theory that big-game hunters came from Asia by hiking across a dry land bridge at the Bering Strait just as the Ice Age waned wasn't really viable? Was it possible, instead, that seafaring people with boats came earlier and by way of the lost world that now lies under water along the North Pacific coast? Witnessing drastic and rapid sea level change for myself piqued my interest in this scenario. So I traveled, asked questions, wrote magazine articles on the subject and learned how, in the face of great obstacles, a few small teams of scientists in Alaska, British Columbia, and California were unraveling the mysteries of those flooded shores. The quest continues, but already the results have begun rewriting the prehistory of our continent.

CHAPTER ONE

Cave of the Bears

"WATCH YOUR HEAD," said paleontologist Tim Heaton, ducking as he led the way down into the fissure in the steep rock face, leaving the sunshine and warmth behind. It was cold, damp, and quiet. Underground, away from the droning of the big gasoline generator, there was only the barest murmur of seeping water. But we could see well enough. Thick electrical cables ran into the cave to a few lights that were strung along one side.

We were on Prince of Wales Island, Alaska, in the summer of 1999. The main chamber was tiny and cramped, only about ten feet wide, with a ceiling so low in places that we had to stoop. The rocky floor was wet and uneven, and the farthest nook was only about forty feet back from the entrance. Kneeling on the muddy floor of the chamber and scraping up dirt with a trowel was a slim young guy with a wispy black beard who turned to say hello. Kei Nozaki, from Japan, was one of the student assistants Heaton had brought up from the University of South Dakota for a season of fieldwork. Like us, Nozaki was wearing a caver's helmet and a rain suit over his clothes.

Off to the right, a second and much narrower tunnel sloped away into the shadows. "The cave branches here," said Heaton with the soft western twang of his native Utah. "This is what we call the 'Seal Passage,' because that's where we found so many of the seal bones. And it actually loops around and comes out at another entrance down the hillside a bit. It's got a tight little keyhole shape." I switched on my headlamp and flashlight and stepped down into the passageway to take a look.

Heaton wasn't exaggerating. The dank and constricted little gut had a symmetrical keyhole cross section all the way along the gentle downward grade to where it jogged out of sight maybe fifty feet away. From knee to head height the walls, glistening with moisture, had the shape of a round tube about three feet in diameter. Below that was a sort of lower extension, a slot barely wide enough to put one rubber boot past the other. This telltale keyhole shape resulted, Heaton explained, because initially, when the water table was high, a steady flow of ground water had dissolved the limestone over thousands of years, routing out a subterranean water pipe, a nearly perfect cylinder, what geologists call a "solution tube." At some later time, the water table dropped and water only flowed along the very bottom of the tube. This eroded the rock straight down from the center and carved out the lower part of the keyhole. Much of the limestone on the island, and elsewhere on the southeastern coast of Alaska, was riddled with similar tubes and passages. Some of the caves penetrate entire mountains and have never been fully explored—a caver's dream.

I figured I would make my way down the length of the seal passage and have a look at the other entrance, but it was so hard to squeeze one foot past the other in the tight slot that I soon had second thoughts. And then I discovered that I couldn't turn around. I was stuck in an awkward crouched position, with cold, slimy stone walls hemming me in on both sides. Being trapped in a mine or buried alive has always been one of my pet phobias. Little fingers of fear crept up my spine, even with Heaton right there, close behind. No, I decided, I did not really need to inspect the other end of that dark, narrow tunnel. I was forced to back my way slowly out and exhaled with relief when I was back in the main chamber.

Heaton chuckled at my discomfort. "I'm used to working under the adverse conditions of caves," he said. "This is my third big excavation on the island." He had done a lot of sport caving as a teenager in Utah—even meeting his wife Julie at a caving club there—and still went caving occasionally just for fun. His other hobby was being a ham radio operator. And, at about forty, he was in terrific shape: tall, lanky, and built like a marathon runner.

For Heaton, the difficulty of working in this accessible, relatively level cave was nothing. "There's one cave here on the island where you have to go headfirst down a tight vertical crawlway, which emerges at the top of a 150 foot cliff." Other caves have underground lakes and flowing rivers. There are tight spots where you have to take off your jacket to squeeze through; if you get stuck, nobody can help you. And if the battery-powered headlamp goes out, everything dissolves into stygian blackness. "I do some pretty hefty caving to get at some of the bones. Cave paleontology is a specialty. Even most paleontologists are not used to it. They'd get claustrophobic crawling through these muddy holes," he laughed.

But it's a handy skill to have in his profession. "There's lots of bones in caves." For one thing, animals can find shelter in caves, and those that hibernate, such as bears, use the caves as dens in winter. A big bonus for paleontologists is the excellent preservation of bones, teeth, antlers—anything made of calcium—in the alkaline environment of a limestone cave. "Bone doesn't preserve too well outside," said Heaton, pointing to the entrance. "You have the temperature and humidity fluctuations, and the acidity of the forest soil, but as you get inside, the bones become better and better preserved, and more common."

The main chamber where Nozaki was digging was called the "Bear Passage." It was a few bear bones Heaton had discovered years earlier, and the totally unexpected information they provided, that initially made this little cave more interesting to him than the many larger and more physically impressive ones on the island.

But it wasn't bear bones that had brought me here from my home in British Columbia; it was human bones and artifacts.

We went back out into the August sunshine, and Nozaki joined us for a break and to warm up from the chill of the cave's interior. Heaton shut down the generator, and we sat to talk among the fuel drums, buckets, hoses, and other clutter of the work site. We were on a south-facing slope in virgin forest of big hemlock and spruce. Heaton took off his helmet and ran his hand through closely cropped red hair. With the sun in his face, he warmed to the quirky tale of how this remote bear den with the official federal designation 49-PET-408 had become one of the most important ancient archaeological sites in North America.

And he might never have even inspected it if it hadn't been for the weather.

In July 1994 Heaton was back on Prince of Wales Island for the fourth time in as many summers. Its caves were proving to be a fossil gold mine, helping him to unravel the complex story of how and when bears first inhabited the Alaska coast. He was one of two paleontologists, plus geologists, biologists, and enthusiastic amateur cavers, who were exploring and mapping the recently discovered cave system with active sponsorship from the local U.S. Forest Service geologist in charge of the vast Tongass Forest District, Jim Baichtal. Major areas of the huge island, the third largest in the United States, were being logged. In fact, it was the logging that had exposed the entrances to many of the caves. But the Forest Service was willing to postpone the clear-cutting in places where caves were being surveyed, and the Feds provided the Tongass cave project with housing, meals, and transportation.

Heaton's plan that year was to spend only a couple of weeks on the island. His target was a high-elevation site called "Bumper Cave," which ran sixty-five feet back from the entrance in sparse subalpine forest high on the flanks of Calder Mountain, a bald rocky dome that loomed over the island's far north end. I had flown past this stark summit early that very morning in the stolid old Beaver float plane that brought me out from Ketchikan to Port Protection, population fifty. There were no roads connecting the settlement to the rest of the island, and no post office, either. Just a dock, a store, a little lodge, a fish plant, and a scattering of houses, all clinging like limpets to the rocky shore of a tiny sheltered cove.

Jim, a chunky, bearded guy, helped me unload my gear from the plane. Like a few other people who were hanging out on the dock, Jim fished a little for salmon with the small commercial fishing boat that was also his home for the summer. Come winter, he drove a taxi in New York City.

To get from Port Protection to 49-PET-408, I hired a lanky fellow named Bud, who was missing a few teeth and had gray hair that reached halfway down his back. He told me that he had moved up from the Lower Forty-Eight in 1972. He made ends meet mainly by cutting lumber for people with a small, portable sawmill, and collected old stone artifacts off the beaches as a hobby. We had to cruise a few miles around a jutting peninsula in an aluminum skiff with a sputtering outboard. Along the way, Bud took a slight detour to show me his favorite spot for finding ancient adzes and daggers made of a greenish stone, presumably by the Tlingit Indians before they first had contact with the Russians and other Europeans in the 1700s. The tide was low and, sure enough, dozens of sharp flakes of green-gray rock were lying among rounded cobbles near some grounded boom logs that were linked together by massive rusty chains. I was skeptical about some of the pieces of stone that he claimed were artifacts, but Bud just smiled knowingly. "Here," he said. "You're going to the cave. Show these to the archaeologists and see what they say." It turned out that he was right.

Bud dropped me off at a gravel beach that was tucked in behind a tangle of kelp. There was a little cluster of brightly colored tents in a sheltered spot above the beach, and just enough room for me to set up mine. But no one else was around; they were all up at the cave, working. From the beach campsite to the cave was a half hour's trudge through dense forest along a steep, muddy trail that climbed more than 400 feet from sea level. For safety, Heaton and his crew had rigged some of the nearly vertical rock faces with knotted ropes. I was panting and sweating by the time I reached the site. It was a hike that I would make morning and evening for the next five days.

Back in 1994, though, Heaton had no intention of hiking that mile-long trail; in fact, it did not exist yet. He and a colleague had assembled the equipment for their expedition up to Bumper Cave, including camping gear and food for at least ten days. They were even bringing along 1000 feet of fire hose, enabling them to run water from a stream to wash and screen the sediment from the cave for any tiny animal bones they might miss at first glance. Kevin Allred, the Alaska-based leader of the cave project, had discovered and mapped Bumper Cave the previous summer, and had regaled Heaton with the juicy details. It was just littered with bear skeletons, he said, probably seven or eight in all, some of them exquisitely preserved. Millennia-old bones from other caves on the island had already overthrown some long-held assumptions about the Ice Age distribution and migrations of both black bears and grizzlies. Heaton was excited by the prospect that Bumper Cave might round out the new and emerging picture.

But his arrival on the island coincided with the long Fourth of July weekend, which grounded the Forest Service helicopter that was to take him and his gear up the mountainside. Baichtal, the geologist, had arranged for him to stay at a Forest Service camp that was the cave project's home base. With comfortable old wooden bunkhouses and separate buildings for cooking and showers, all overlooking a sheltered arm of the sea, this was not exactly a hardship post. But then the coastal weather closed in, and the helicopter could not fly. Day after rainy and foggy day, Heaton was forced to wait, cooling his heels at the camp, a frustrating thing for a guy who hates to waste a moment.

Waking one morning to see the trees shrouded yet again in mist, he and Allred decided to take a side trip to another cave on the island's extreme northwestern tip that Allred had visited the year before. Allred had heard about it from a Forest Service surveying crew, who first noticed the dark opening in the hillside when they hiked through to plot out future logging roads. The cave's ceiling was so low that the initial designation 49-PET-408 eventually gave way to the moniker "On Your Knees Cave."

Allred and two colleagues had crawled through with their compass, inclinometer, and tape measure back in 1993. Later, they made a scale drawing for the cave project's files, and that might have been the end of it, except that they also noticed some animal bones, in particular the well preserved skeleton of a river otter. When he heard about it from Allred the following summer, Heaton wondered what else the cave might hold.

With the helicopter grounded, Heaton and Allred drove to the end of the main logging road and hiked through the woods out onto a peninsula called "Protection Head." The cave didn't look like much. Inside, it was as Allred had described it. There was the river otter, which was interesting. *Gee, imagine an otter coming all this way up from the sea*, Heaton said to himself. And there were a few larger bones lying exposed right on the surface. One of them, he thought, was the femur of a grizzly bear. This wasn't a surprise to Heaton. Grizzlies did not inhabit the island in modern times, although black bears did. And though it used to be thought that grizzlies never had, Heaton had been finding the bones of both grizzlies and black bears in many of the caves for years now.

Heaton put the visible bones into plastic bags, labeled them, noted their locations in his notebook, and returned to the Forest Service camp. Eventually the weather cleared, the helicopter took him and his partner up to Bumper Cave, and they collected a wealth of bear bones, which later proved to be from 7,200 to 11,600 years old. It was a highly successful, if short, season. As for On Your Knees Cave, Heaton wrote it off as having much less paleontological interest than many of the other caves on the island. All he had found, after all, were the otter skeleton and a few more bear bones.

His verdict would change drastically a few months later. He had sent off a minute sample of the grizzly bear femur to a specialized laboratory for carbon dating. The lab isolated perhaps a thousandth of a gram of bone protein, called collagen, and sent it on to a nuclear laboratory equipped with a high-energy linear accelerator linked to a mass spectrometer. By what is called accelerator mass spectrometry (or AMS) dating, individual ions of the radioactive carbon isotope, carbon 14, were counted. The results went back to the first lab for analysis.

Back in South Dakota, Heaton eventually received a carbon 14 date of 35,000 years for the bear femur, or three times as old as the oldest bear bone previously found on the island. This placed it well *before* the peak of the last glaciation. Bones from the other caves already showed that bears had inhabited the island in the final, waning years of the glaciation, around 12,300 years ago. Was it possible that they had survived there right through the most recent advance and retreat of the great ramparts of ice, from about 25,000 to 10,000 years ago? If so, the area was a glacial refuge, a sizable haven of life on the outer edge of what was for many millennia a largely sterile, frigid, and icebound coast.

These findings ran entirely contrary to what biologists and geologists had been saying. Most thought that the entire coast of southeastern Alaska, including the islands like Prince of Wales, had been smothered by ice, leaving no such refuges. In 1965, biologist David Klein undertook a study of mammal distributions in the islands of southeast Alaska and had proclaimed unequivocally: "During the [most recent] glaciation the present land areas of the coastal regions . . . were virtually completely overridden by ice. The now existing flora and fauna of the region have presumably become established in the 10,000 years since the recession of the ice." For decades, this remained the authoritative view.

Recently, though, a few scientists had begun to wonder. Even Heaton's earlier finds of bears dating to the very end of the Ice Age, from 10,000 to over 12,000 years ago, were a significant clue. Jim Baichtal had told the journal *Science*, "If bears were living here, then chances are pretty good that we were not overridden by a blanket of ice as the textbooks have been telling us."

And there was other evidence that pointed to the existence of Ice Age refuges. Grizzlies still inhabited islands farther north along the coast of Alaska. Several studies of their DNA by Gerald Shields and his colleagues at the University of Alaska showed that these grizzlies were more closely related to polar bears than to any other currently living bears, including other grizzlies that still inhabited the adjacent mainland of Alaska. And this affinity was despite the fact that the intervening channels were not too wide for the bears to swim across.

This suggested that the offshore grizzlies had spent a very long time evolving in genetic isolation. Polar bears, in turn, were thought to have evolved from brown bears, the genus to which grizzlies belong, probably on the Siberian coast. The most likely explanation for this overall pattern was that the offshore Alaska grizzlies had been cut off from contact with other grizzlies during the last glaciation in a coastal refuge, or *refugium*, as the scientists like to call it. And it would have to be quite large to support a separate breeding population of bears. Now, thought Heaton, the bones from On Your Knees Cave might be telling a similar story.

Heaton had already tossed ideas back and forth with a few geologists and archaeologists who thought there might have been an entire network of such offshore refuges along the North Pacific coast. He realized that the northwestern tip of Prince of Wales Island had not likely been a tiny, isolated pocket of ice-free land. To the west and north of Prince of Wales Island—and therefore farther away from the mainland coast with its massive ice sheet—were other very sizable islands. If the tip of Prince of Wales had been beyond the reach of the ice, major parts of those other islands would probably have been ice-free as well.

Although the Cordilleran ice sheet may have been a mile or more thick where it blanketed the higher mountains of the mainland coast, at its outer edges to the west it would have thinned significantly, and as it advanced it would have split into separate glacial flows that followed the paths of least resistance. So it would be wrong to picture a monolithic wall of ice fronting the sea, like the vertical blue cliffs of an ice shelf on the coast of Antarctica. What reached out among today's coastal islands to the then-lower sea level on the continental shelf would have been massive snouts of grimy gray ice, oozing their way downhill under the force of gravity like stiff molasses. Where there were mountains or high ridges on the islands, these frozen rivers, propelled by the weight of ice behind them at higher elevation on the mainland shore, would have followed the deeper channels between the islands to grind opportunistic paths around the obstructions. And, because of lowered Ice Age sea level, there would also have been land beyond those islands that blocked the glacial flow, out on the gently sloping continental shelf. This land is now under water. At the peak of the glaciation, though, it would have been dry and quite habitable.

Tom Ager of the U.S. Geological Survey (USGS) has been systematically tracing the glacial flows through the archipelago of southeast Alaska and pinpointing these formerly exposed areas of land on bathymetric charts. To the west of Dall, Coronation, and Baranof Islands, says Ager, there were a number of distinct glacial refuges, a few of them hundreds of square miles in size.

To Tim Heaton, all of this suggested that along with the bears, people might have lived in the area as well. If these people had camped, fished, and hunted on the shores of an extensive glacial refuge, much of their prime territory would be under water today. But the cave could have acted as a magnet to draw them inland and uphill to hunt the denning bears, so it might offer the best chance to find traces of these early coastal dwellers. *If*, that is, people were actually on Prince of Wales or neighboring islands back in Ice Age times. Heaton was excited at the prospect of making a discovery that could overturn some of the most strongly held and long-established ideas about North American prehistory. He could hardly wait for the next summer, to get back to On Your Knees Cave and investigate further.

It would be two years before the human story of On Your Knees Cave began to unfold. And when it did, it went a long way to unravel the secrets of the late Ice Age coastal world.

But not all the pieces of the puzzle would come from the efforts of a lone paleontologist gathering bones in an Alaskan cave. Some would be dredged from the sea floor by a research ship in the Queen Charlotte Islands of British Columbia, a few hundred miles to the south. The bones of an extinct mouse species on an island off California would offer mute testimony about when an Ice Age woman died in a swampy canyon. Important information would come from a geologist who determined the length of time that a line of erratic boulders hundreds of miles long had been exposed to the sky. Other pieces of evidence, including razor-sharp microblades of volcanic glass, would be unearthed by young Tlingit and Haida Indians, the likely descendants of the very ancient people who left those artifacts behind.

It was a heady experience for those involved in the search. They knew that what they might find could revolutionize our most fundamental self-image. For we have been accustomed to thinking of ourselves as a species in terrestrial terms: evolving in the savanna of Africa; hunkering in caves in Europe; gradually spreading overland through Asia; and finally trekking dry-shod across a land bridge at the Bering Strait into the Americas while preying upon big Ice Age animals. But if the scientists on the Pacific coast were right, we also became bold seafarers at a very early date, maritime people who built boats and braved the stormy and icebound shores of the North Pacific. And we lived not just from hunting mammoths and huge bison, but also from spearing sea mammals, from fishing, and from gathering shellfish and seaweed. This would be quite a different picture, yet one that is just as heroic and compelling.

CHAPTER TWO

Northern Connection

TOM GREENE JR., a muscular Haida tree faller, had worked closely with professional archaeologists on seaside excavations, on ships, and even as a diver on an underwater dig in the Queen Charlotte Islands. Greene looked bemused one day when discussion turned to precisely when people first arrived on the Northwest Coast. Was it 12,000 years ago, or perhaps much earlier? "All this number crunching that I hear," said Greene. "If you ask me how long we've lived here on the Charlottes, I'll just say, *since day one.* There's nothing in our history that says we were beaten or chased away from anywhere else, and we didn't chase anyone away to get here. We *were* here."

Most Native American oral histories include stories of creation that are set in their traditional territories, while relatively few involve migrations from distant places. Take the Haida, who live on both the Queen Charlotte Islands of British Columbia, Canada, and on nearby Prince of Wales Island, Alaska. Haida tradition tells of ancient times when they could walk from one island to the next over open grassy plains, before the arrival of the first tree. And it tells of a great flood.

I first heard the flood story from Captain Gold, a Haida elder who has coauthored several archaeology papers. He also spearheaded a campaign to preserve Sgan Gwaii, a unique abandoned village full of carved cedar poles, a place possessed by spirits, where heraldic wooden figures loom over a grassy clearing and defy the petty passage of generations.

"There's one old story about a village that was very close to the shore when sea levels were much lower," he told me as we sat on the foredeck of a research ship. A strange woman appeared on the beach one day, wearing fur from an unknown animal. Children from the village harassed her, pulling off the fur and revealing that her backbone was encrusted with barnacles. When they laughed at her, the elders warned them to stop, lest some harm befall them all. But the children persisted. "And sure enough, she turned out to be a supernatural being, Flood-Tide Woman. She raised her skirt, moved back behind the beach, sat down, and the tide came up to follow her. The children kept mocking her, so she kept doing it. Pretty soon she was up beyond the village, then up into the forest, then up into the mountains. The tide kept rising, and soon everything was covered with water and everyone drowned. Except for a few people who were lucky enough to be in canoes. Once the sea began to drop, those survivors repopulated Haida Gwaii," which is what the Haida call their home islands.

The Haida tradition evokes strong echoes of late Ice Age conditions. But as is typical for Native American oral history, there is no tale of coming from elsewhere.

The Europeans who arrived in the wake of Columbus had quite different ideas. Explorers, conquistadors, clerics, and early settlers became curious about the origins of Native Americans almost as soon as they encountered the incredibly diverse peoples and advanced civilizations of the American hemisphere. They wondered who these aboriginal Americans were, how they fit into the larger human picture, where they might have come from, and how long ago.

An extremely prescient early guess was made in 1590 by the Jesuit missionary Fray Jose de Acosta, who thought small bands of hunters from Asia likely came to America via a land bridge or narrow strait in high northern latitudes. What is so remarkable here is that in Acosta's time Europeans could only speculate about whether the northwest-trending Pacific coast of North America came even close to linking up with Asia. The consensus among sixteenth-century mariners and geographers, based on little more than the hope that this might provide a shortcut from Europe to the riches of China, was that a broad strait, or "Northwest Passage," separated Asia and North America roughly at the latitude and longitude of the Gulf of Alaska. The actual geography of the far North Pacific, with its very narrow Bering Strait located much farther to the west and north, was not revealed until the eighteenth century, first somewhat vaguely by the Russian-sponsored voyages of Vitus Bering in 1728 and 1741 and then more precisely in 1778, when the third great voyage of Britain's James Cook followed the Alaska coast and crossed the Bering Strait to Siberia.

During these centuries the nearly universal view was that the Indians had come to America from somewhere else. Why? At first, the reasoning was based on religion.

According to Scripture, all human beings were descended from Adam and Eve, which meant that Indians must have migrated to their current homelands after the Deluge, presumably from Asia Minor, where Noah's Ark had landed. Some thought American Indians were one of the lost tribes of Israel. Others, harking back to Plato, saw a link to the lost continent of Atlantis. In the early eighteenth century, Christian doctrine still heavily influenced the speculation. The prominent Boston Puritan clergyman Cotton Mather wrote that "probably the Devil decoyed these miserable savages hither, in hopes that the gospel of the Lord Jesus Christ would never come here to destroy or disturb his absolute empire over them."

By the end of the eighteenth century, the very different approach of Thomas Jefferson reflected the emerging spirit of scientific inquiry that was a hallmark of the Enlightenment. There were large burial mounds in the Ohio and Mississippi valleys, and smaller ones near Jefferson's home in Virginia. These "barrows" might have been put there by the Indians or their ancestors, although some white Americans initially thought they were earthworks erected in the 1500s by the Spaniard Ferdinand De Soto to defend against Indian attack. Jefferson rejected such speculation and wrote that "we must wait with patience till more facts are collected." But then he went and carefully dug trenches through a mound near his estate and studied the bones, an act that has been called "the first scientific excavation in the history of archaeology." And when he learned of the recently discovered bones and teeth of extinct elephantlike mastodons, he quizzed many Indians to find out what they might know about such animals.

Jefferson had known the local Indians since childhood and had a highly sympathetic, even romanticized, view of them. At eighteen, he visited one of the great chiefs of the Cherokee, and he continued to study Indian cultures throughout his life. He collected so many of their artifacts that the entryway of his mansion, dubbed the Indian Hall, was lined with bows and arrows, peace pipes, wampum belts, and carved stone figures. He also assembled the vocabularies of thirty Indian languages. Based on the sound of these languages, the physical appearance of living aboriginal Americans, and the narrowness of the Bering Strait, Jefferson concluded that they and northern Asians probably had a common origin.

But how old might this link to Asia be? Most early guesses were that the Indians had arrived only a few thousand years ago. One reason for this relatively short time frame was that, according to Christian theology, the entire world had only existed for around 6000 years. In the 1650s, Irish Archbishop James Ussher had carefully gone through the Bible, noting who begat whom and counting backward through the generations. Based on his calculations, he proclaimed that the year of Creation was precisely 4004 B.C.

Benjamin Smith Barton, later a professor of medicine, natural history, and botany at the University of Pennsylvania, was a contemporary of Jefferson's and the first American with scientific training to delve into the origins and antiquity of the Indians. Barton traveled to Ohio, studied the burial mounds, and likewise thought that Indians and Asians probably shared a common heritage. He further concluded that the great diversity of the native languages could not have emerged within the span of only a few thousand years. Native Americans, he thought, must have been separated from their Asian relatives for such a long period that Archbishop Ussher's chronology could not be correct. This approach—using the diversity of native languages in the Americas as a yardstick to estimate the passage of time—is still being applied by some linguists today. But nobody in the eighteenth or early nineteenth century knew even roughly what kind of time frame was involved.

In the late twentieth century, Tim Heaton could send his samples of bear bones off to a laboratory and get back a figure of around 35,000 years. He could relate this to the ten thousand years that have passed since the end of the Ice Age, or to the several million years since our hominid forebears first evolved, or to the several billion years that the Earth has existed. But Thomas Jefferson had no idea at all how old the world was, let alone how long the Indians had been in America.

It was geology that would finally revolutionize the Western view of time. In the eighteenth century, James Hutton, a Scot, noticed a number of "unconformities" in the sedimentary rock of cliff faces, places where a neat sequence of horizontal rock strata lay atop another set of parallel strata that were nearly vertical. Hutton inferred that these contrasting formations had to represent quite distinct periods of sedimentation. He recognized that sedimentary rock was formed when land surfaces were eroded by wind and water, carried down to the sea by rivers and deposited on the sea bottom, eventually forming solid rock with horizontal layering. Then some immense force such as volcanic activity must have uplifted the strata, broken them, and tilted them to the vertical position. After that, another period of erosion and sedimentation—once again, under water—had apparently occurred to lay down the strata closer to the top of the cliff. Hutton likened the Earth to a machine that cranked its way through repeated cycles of change, with "no vestige of a beginning, no prospect of an end."

Charles Lyell, a fellow Scot, wrote that Hutton's theory assumed an "immensity of past time" and provoked "a painful sense of our incapacity to conceive of a plan of such infinite extent." But where Hutton saw gradual processes alternating with intermittent cataclysms, Lyell proposed in 1830 that extremely slow but inexorable processes could, given sufficient time, account for virtually all geological phenomena. His approach required a vast, almost unimaginably long earthly chronology, at least in the many millions of years. Lyell had discovered what American science writer John McPhee has called "deep time." Oddly enough, that same decade German astronomer F. W. Bessel discovered "deep space." His measurements showed that even the nearest stars were not just millions of miles away, but at least *millions* of millions. It was an age of mind-bending intellectual upheaval. And there was more to come.

In the mid-1830s Swiss geologist and naturalist Louis Agassiz (later a professor at Harvard) noticed that valleys in the Alps had distinctive shapes. Where glaciers were seen to be at work, the valleys were broad, almost flat-bottomed, and had a U shape. Other valleys, cut by the action of flowing water, were narrow and V-shaped. He also found that many U-shaped valleys in the Alps and elsewhere currently had no ice associated with them. Agassiz argued that at one time glaciers must have flowed down these valleys as well and carved them out to that telltale contour. The existence of long-vanished glaciers would also explain how huge boulders made of nonlocal rock, called "erratics," could end up strewn across the landscape, how some exposed rock surfaces came to have a polished look to them, and how long sets of straight lines became scratched in the surface of bedrock, apparently by the movement of the glacial ice. Agassiz proposed the radical notion that during a great glacial epoch in the distant past a vast northern ice sheet had covered much of Europe. At first, most geologists vehemently rejected his theory.

In 1840 Agassiz was invited to visit northern England and Scotland to examine some surface geological deposits that were clearly quite different from the underlying rock. The young and as yet little-known Charles Darwin had studied these same formations at Glen Roy, Scotland, and claimed in a scientific paper that they were former marine beaches. Agassiz concluded instead that they were glacial moraines, distinctive lines of earth and rock that were deposited across the front of the glacier and left there when it melted. Similar moraines could be seen stretching across other valleys in England, Scotland, and Ireland. Agassiz wrote an article for a Scottish newspaper arguing for the "former existence of glaciers" in the highlands of Scotland and presented a paper at the prestigious Geological Society in London.

His facts and logic persuaded Britain's top geologists, including Charles Lyell. Darwin initially rejected Agassiz's view, mainly from stubborn pride, and wanted to muster a last-ditch defense of his own shaky reasoning. But Lyell, his friend and sponsor at the Geological Society, advised him to accept the inevitable and retract his paper. "You'll have to eventually," Lyell told his protégé with affection, "the sooner the better. In art or literature one never has to admit that he has been wrong. In science, yes. . . . We must dare to explore and to theorize from our findings, to learn as we go along." Darwin withdrew his paper, and the existence of ice ages gained credibility.

Meanwhile, in America, physical anthropologist Samuel Morton was putting his calipers to skulls taken from the Ohio burial mounds and comparing them with skull measurements he had taken of modern Indians. In his book *Crania Americana*, Morton concluded that the ancient and modern Indians all belonged to a single "Mongolian" race. (Morton's conclusion is now considered wrong, and many scientists reject the very notion of "race" as a meaningful concept, preferring to speak of diverse human populations with varying degrees of affinity to each other. But physical anthropologists still measure facial and other features and make comparisons to try to determine ancient origins. The most notable current example is the work of C. Loring Brace of the University of Michigan, who believes the earliest Native Americans may be derived from the ancient Jomon people of Japan.) How old this genealogical connection to Asia might be, though, was still a mystery.

The genealogical connection between Native Americans and Asians became better established in the early twentieth century. Physical anthropologist Ales Hrdlicka of the Smithsonian Institution found that the incisors of American Indians (the biting teeth at the front of the mouth) were distinctly scooped out on the back. This pattern, which serves no apparent purpose, was also true of most Asian teeth, but much less common among Europeans and Africans. He wrote in a massive 1907 report, *Skeletal Remains Suggesting or Attributed to Early Man in North America*, that America had been peopled by immigration from Asia, probably right at the tail end of the Ice Age or shortly after the last glaciation. The biological link across the north seemed to be strong.

A further hint, though, came from biogeography, the study of the distribution and dispersal of species. The tropical animals of the New and Old Worlds had relatively little in common. New World and Old World monkeys, for example, were so different that biologists classified them as belonging to entirely different families, which reflected what scientists now know were the many millions of years that they have evolved in isolation from each other since South America and Africa drifted apart. In the northern temperate latitudes, though, the differences among animals were much less pronounced. And in the Arctic regions the animal species of Eurasia and North America were almost identical. This suggested that these great land masses had been joined at some relatively recent time by a land connection in the far north.

More concrete evidence appeared in the early 1890s when a decades-long dispute between Britain and the United States came to a head. The Pribilof Islands, which lie in the Bering Sea, had vast colonies of valuable fur seals. The United States claimed sovereignty as part of its ownership of Alaska, allowing American hunters to take 100,000 each year. When Canadian sealing ships intruded on the orgy of slaughter, the United States seized a number of them, even though they were hunting beyond the three-mile limit. Britain still wielded the not-yet-tattered reins of empire on Canada's behalf. A commission was established to study and arbitrate the so-called "Bering Sea Controversy." Britain named as one of its commissioners George M. Dawson, an experienced surveyor and assistant director of the Geological Survey of Canada (GSC). While Dawson cruised the Bering Sea making an inventory of the local marine resources, his crew took extensive soundings with the lead line. These showed that water depth in the Bering Sea was consistently less than 100 fathoms (600 feet). The sea bottom, Dawson wrote, was part of the "continental plateau region as distinct from that of the ocean basins proper." He argued that "in later geologic times more than once and perhaps during prolonged periods [there existed] a wide terrestrial plain connecting North America and Asia."

Around the same time, separate expeditions found the fossil remains of extinct mammoth elephants on the islands of Unalaska (in the Aleutians) and on the Pribilof Islands (in the Bering Sea). Both locations were surrounded by shallow water, virtually proving that much of the continental shelf in the Bering Sea area had at least once been dry land that connected these islands to the mainlands of Siberia and Alaska.

And so the idea of an ancient land bridge was firmly established. But what conditions or geological process accounted for it, and precisely when had it existed? Geologists knew that drastic episodes of uplift, subsidence, and sea level change had occurred all over the world, but they did not understand the mechanisms that caused these vertical movements. The theory of plate tectonics, which explains, for example, how the Indian subcontinent is even now colliding with the Asian continent and pushing up the Himalayas, was only developed in the 1960s. Geologists, therefore, saw no causal connection between the timing of the last glaciation and the existence of the land bridge.

It was the sheer vastness and thickness of the ice sheets, and thus the amount of water they held, that provided the final clue to why and when the land bridge had existed. By the late 1920s, geologists and glaciologists had made great progress in mapping out the terminal moraines left by glaciers in North America and thereby delineating the margins of the great ice sheets of the last glaciation. They also measured the elevation at which glacial erratics had been deposited on mountain slopes, which gave a rough estimate of the thickness of the ice sheets. The western or Cordilleran ice sheet, which extended from the Rockies to the Pacific coast, had filled all the valleys and reached about 8000 feet high on the mountain ranges. Above this level rose only bare and isolated peaks, spectral ice-fringed islands in the sky that scientists call "nunataks."

The eastern or Laurentide ice sheet, which extended from east of the Rockies to the Atlantic, was quite different. It had several domes of ice centered over low-lying and mountainless terrain, one in the Keewatin area west of Hudson Bay and another near north-central Quebec or Labrador. The thickness was mainly dependent on snow buildup in areas of high precipitation. The weight of these raised domes exerted such powerful outward force that ice at the margins was in some regions of rising ground actually pushed *uphill* for hundreds of miles. Geologists found erratic boulders from the centrally located Canadian Shield at elevations up to 4350 feet in southern Alberta and northern Montana. From this kind of evidence, glaciologist Ernst Antevs estimated in 1928 that the Keewatin dome was 18,000 feet thick and the Labrador one about 14,000 feet.

Canadian geologist W. A. Johnston of the GSC reviewed the information on the extent and thickness of the ice sheets and related it for the first time to the existence and timing of a link to Asia that people could have used to migrate. Clearly, enough water had been locked up in the ice sheets that it would have had a major impact on world sea level. Johnston concluded in 1933 that during the last glaciation "the general level of the sea must have been lower owing to the accumulation of ice on the land. The amount of lowering is generally estimated to have been at least 180 feet, so that a land bridge probably existed during the height of the last glaciation." In fact, 180 feet was a significant underestimate, and the term "land bridge" conveys the false image of a narrow connection like the Isthmus of Panama. Because the Bering Sea to the south and Chukchi Sea to the north are very shallow, with 350 to 400 feet of water removed, the land bridge was really a broad subcontinent about a thousand miles wide.

Swedish geologist Eric Hulten christened this vast, largely lowland, and relatively dry and ice-free arctic region that had linked the continents "Beringia," and suggested that it had lain exposed during a number of glacial episodes, including the most recent, which geologists in North America called the "late Wisconsin glaciation," and Europeans called the "Würm." And he argued that it was the route by which early hunter-gatherers had reached the New World. Beringia was firmly established as the gateway to the Americas.

But even if people had been able to cross from Siberia into Alaska when sea level was low, they might not have been able to move farther south until after the Ice Age. When geologists mapped the northern ice sheets, they seemed to cover most of Canada from coast to coast, along with nearly all of the southeastern coast of Beringia, creating a nearly impassable barrier. Johnston, however, proposed that as the Ice Age waned, an ice-free corridor opened up between the Laurentide and Cordilleran ice sheets and permitted migration through central Canada from the land bridge area to the middle latitudes of North America, essentially today's continental United States. But no one was sure.

What *was* almost certain was that Ice Age people could not have trekked thousands of miles across glaciers devoid of plants and animals. So the exact location of the ice—the extent of its reach and the timing of its retreat—seemed to hold the key to understanding when people could have reached North America's temperate latitudes.

Before heading for Alaska in 1999, I pored over the most authoritative study of the late Wisconsin glaciation that I could find. A huge, oversize tome that had been published in 1981, *The Last Great Ice Sheets* was chock-full of large and detailed maps, all well footnoted to indicate which peer-reviewed papers from geological journals were the basis for its conclusions. But it was inconclusive on the important issue of whether or not an ice-free corridor existed where the two ice sheets either met, or *nearly* met, to the east of the Rockies. The editors presented two quite different maps of the ice margins in this region and were frank in admitting that the scientific data on which they were based were ambiguous. One map showed a gap between the ice sheets in late glacial times, an ice-free corridor about two thousand miles long extending from the northern Yukon Territory to northern Montana. The other had the Cordilleran and Laurentide ice sheets coalescing east of the Rocky Mountains, and therefore blocking any migration route until the very end of the glaciation about 10,000 to 11,000 years ago. Geologists simply did not have enough information at that time to be sure.

As for where the ice reached along the Pacific coast, though, the maps were strangely definitive, especially considering how little actual research had been done on the question at that time. The maps showed the Cordilleran ice sheet extending right out from the mainland to cover *all* the islands. This included the largest, Vancouver Island, Prince of Wales Island, and Kodiak Island, as well as those that lay farthest off the mainland shore, the Queen Charlottes in British Columbia and Dall and Baranof Islands in southeast Alaska. For at least the next fifteen years, most archaeologists and paleontologists had simply assumed that there could have been no animal or human life worth looking for out on these islands. USGS geologist Tom Ager, whose recent research has shown just how large the ice-free offshore areas in southeast Alaska actually were, explained to me that "once a glacial extent map gets published, it tends to be adopted by other authors, and it gets entrenched in the literature to the point that the maps acquire a credibility that is unjustified." Scientists, like anyone else, can get lazy and lackadaisical, and remote islands are not the easiest places to do field research. There was little incentive to challenge the orthodox view.

Initially, Tim Heaton had also largely accepted the prevailing opinion. When he found the 35,000-year-old grizzly bear bone, though, he realized that the Pacific coast might *not* have been entirely smothered in ice; grizzlies might have survived in glacial refuges. It was still very much an open question, but he was sympathetic to the problems faced by geologists. The North Pacific coast was so long and convoluted in shape, and so intricate in its geology, that much of it had never been properly studied. And even where geologists had tried, it was not easy to unravel the sequence of events. Where glaciers extended out onto the continental shelf, their terminal moraines were now underwater and difficult to identify even with the help of modern seismic imaging technology. Other evidence of glacial extent, such as erratic boulders and telltale scratches on bedrock, could easily be hidden in the dense cover of coastal rain forests.

Heaton explained how tricky the whole exercise could be. It was the first day of my visit to Prince of Wales Island, and we sat outside the cave in the sun. "It was thought that this whole area was overridden with ice during the last ice age," he said, pointing down into the little valley below, which had been scoured out by a river of moving ice. Even an Alaska geologist who had coauthored a scientific paper with Heaton took that position. "Tom Hamilton was up here for a couple of years of work. He had different ideas from the rest of us, although I think he kind of came around to our way of thinking after a while. One problem is that it's really hard to tell *when* glaciation occurred. Tom was looking at the big picture of glacial valleys. And, you know, it's true. Certainly there's a lot of evidence that glaciers have overridden this area. But when? Was it really the late Wisconsin, or the early Wisconsin, or some even earlier period?"

In other words, after a series of glaciations it was extremely hard to distinguish the effects of one glacial advance from another. Lacking conclusive evidence, the mainstream of anthropology simply accepted the supposed authority of what the geologists were telling them, that the Cordilleran ice sheet had bulldozed its way out onto the continental shelf, leaving no place for animals or people to live. And if that were true, then certainly there would have been no place for people to stop, rest, and find food if they were traveling by boat, which most archaeologists thought was highly unlikely to begin with.

Decades of gradually accumulating archaeological evidence conspired to support this view. Right through the 1960s and 1970s there were very few solidly dated archaeological sites on the Pacific coast that were older than about 8000 years, whereas in the interior of the continent there was abundant proof of human occupation 10,000 and more years ago. Archaeologist Jim Dixon, who would collaborate with Heaton at On Your Knees Cave, recalled his own education in the 1960s. “When I was a student,” he told me, “we were taught that coastal and maritime adaptations, at least in North America, didn’t really happen until around 4000 to 5000 years ago.” At that time there were few known coastal sites any older.

So the well established idea that people first came to the Americas by walking over a dry land bridge, combined with glacial maps that showed the coast buried in ice and an inland migration corridor *possibly* open, created a strong bias toward terrestrial migration by the continent’s first inhabitants. As with any self-negating prophecy, if people don’t believe in something, they’re not likely to look hard for evidence to contradict their opinion. In fact, they may resist it, as even an intellectual giant like Charles Darwin was tempted to do. “Archaeologists can get dogmatic,” Dixon went on. “Their careers are often dedicated to a certain hypothesis or position, and when it’s threatened, they see their careers and entire life’s work going down the tube. So they get defensive. It’s human nature.”

Thus, the small world of experts who studied the first peopling of the Americas focused its efforts and attention almost exclusively on the interior. The few scientists who believed in early coastal migration were fighting an uphill battle, both for funding to support their research and to be taken seriously when they presented evidence that supported their position.

What they needed was a breakthrough: decisive proof that offshore glacial refuges existed and that people had lived out on those remote coastal areas back in Ice Age times. As they pursued their search in the mid-1990s, they did not realize how close they were to having that proof in their grasp.

CHAPTER THREE

Important Artifacts Found

TIM HEATON SHOOK his head in amazement. "No one had even *looked* for fossils" of glacial age in southeast Alaska, he said, at least not until the Prince of Wales Island cave project began. Because the biologists and geologists were so convinced that the whole coastal area had been locked under a forbidding bastion of ice, paleontologists did not consider the island a very likely spot to find traces of life extending back into the peak glaciation years.

It was still my first day visiting the cave site. The generator that powered the lights inside the cave was off, and it was blissfully quiet where we sat on the sunny hillside nibbling our lunch, miles from the nearest house or road. Below us, the dense forest canopy dropped off sharply, while above the cave a sheer cliff was topped by steep woods that soared out of view. Heaton wanted me to understand just how hard it was to unravel the glacial history of the area, and he was eager to share the excitement he had felt as the first tantalizing evidence of ancient human presence here came to light.

A lot of sleuthing was required to figure out precisely where the glaciers had flowed. The dated animal bones found in the various caves on the island provided a few points in space and time where the evidence was solid and convincing. More recently, geologists had begun taking pollen samples from bogs, lakes, and peat lands on the island as well, to fill gaps in the picture. But the overall scenario of glacial advance and retreat was still not an easy puzzle to decipher. Grinding rivers of ice could leave a variety of footprints; a later glaciation often obliterated the traces of previous ones; and the timing of different glaciations was especially tricky to determine.

I tried to visualize our lunchtime spot as it might have been 12,000 to 15,000 years ago, before the trees appeared. It would have been an outcropping of mainly barren bedrock, with perhaps a thin veneer of soil and some tundra plants. A vast front of ice would be slowly melting and retreating eastward, possibly just over the hill behind us. I wondered whether little accessible pockets of life, such as the cave's surroundings, could have existed even in areas that were otherwise heavily impacted by ice. And did it matter which side of a hill you looked at? Could flowing ice scour one part of an area like this—such as the valley bottom—and yet spare other places nearby, for instance up here on this ridge?

"If a glacier moved through a valley," Heaton said, "it would pretty much hit both sides. But this cave is on a south-facing slope, and that's probably very important. The other side of the canyon here may have been covered with snow and ice much of the year—not glaciers, where you have huge flowing masses, but ice fields. This side might have been a much more lush area, with more plants and much more hospitable." It would have been a lot more attractive to animal life and perhaps to people as well. "El Capitan Cave is also on a south-facing slope. So that's something of a pattern that has emerged."

I had already heard and read about El Capitan, the island's most famous cave and the first one that Heaton and his group had extensively explored. Truly spectacular, it is the largest known cave in Alaska, with more than two miles of passageways on three levels. The more Heaton talked about it, the more my curiosity was piqued, enough in fact for me to change my plans. Following my five days at On Your Knees Cave, but before leaving Prince of Wales Island, I arranged to spend an extra day and night at a local fishing lodge in Port Protection. And I asked the lodge owner, Bob Gray, to take me to El Capitan, which meant going by boat to where he kept his truck, and then driving south for an hour on logging roads. The cave overlooked a winding channel on the island's heavily wooded western side. Unfortunately, it had become such a popular place to visit that the Forest Service installed a locked gate at a narrow spot about a hundred yards in from the main entryway.

The geologist who sometimes took people in by appointment was not available the day we went, but there was a knowledgeable caretaker, Joe Neynaber, who lived in a mobile home at the site. He lent us helmets and flashlights, so we could at least peek in at the first section of the cave. The contrast with tiny On Your Knees Cave couldn't have been greater. At El Capitan we walked erect through an echoing gallery with a high vaulted ceiling, like a medieval cloister. Narrow tubular shafts, or "chimneys," shot straight up from that ceiling and disappeared above us into the gloom. I had a hard time imagining how, even with technical climbing gear, the cavers ever made it up such constricted passages to reach the higher parts of the cave, but that's what they had done. We skirted gaping, water-carved pits in the rocky floor, where you could easily fall thirty or forty feet to your death. Shining my flashlight through the locked gate, I could make out the beginning of a broad passage that meandered far off into the bowels of the earth.

Neynaber had a map that showed the cave extending thousands of feet back into towering El Capitan peak. "There's a river running right through the cave," he said. "Some places at the back can hardly be explored, because they're cut off by water. The whole mountain is just honeycombed." The grandest chamber, he said, was about 230 feet long, 120 feet wide, and many stories high, just about big enough for a softball field. The entire place was huge and strange and spooky. It was full of bats, and albino shrimp lived in the water.

Even in ancient times there had been odd goings-on at El Capitan. In the 1980s, a geologist noticed some unusual bats that hibernated in the cave, and then a few obsidian spear points. He tipped off archaeologists, who poked around the artifacts and discovered an altarlike shelf containing the ceremonial burial of a river otter hundreds of yards in from the entrance. The animal had been wrapped in cloth made from cedar bark, and near it were little piles of ashes from the torches that had been used to light the way into such deep recesses. No one could explain the significance of the burial. The date on the ashes was about 3200 years. The cave was also full of fragile stalactite and stalagmite formations, and the walls oozed with droopy layers of flowstone. Given the sheer size, difficult access to certain parts of the cave, and the musty odor of mystery, it was no surprise that El Capitan thrilled serious local cavers and became their first target for detailed exploration.

Tim Heaton and Kevin Allred, the same caver who would take Heaton for his first look at On Your Knees Cave, had gone spelunking together in Utah when they were teenagers. Eventually, Allred and his wife Carlene moved up to Haines, Alaska. They were both dedicated amateur cavers. Heaton credited Kevin Allred with being "the one who got the whole caving expedition started up here. When they heard there was limestone on Prince of Wales Island, they decided to take a family vacation to check it out. I think that was around 1986. They started mapping out El Capitan Cave, and the Forest Service got interested in what they were doing." The Forest Service agreed to provide food and accommodations and invited other cavers up from the Lower Forty-Eight. "So it became an annual summer expedition," the Tongass Cave Project, named for the region's Tongass Forest District.

From the start, Heaton asked the Allreds to keep an eye out for fossils in El Capitan. At first, they encountered mud and moisture that made Heaton figure there would be poor preservation of animal bones. Then he received a letter that changed everything. Kevin Allred had discovered a route up some of those tongue-bitingly vertiginous chimneys into a high remote passage. "The first thing Kevin found there was a whole bear skeleton, just laid out in beautiful, perfect condition," Heaton chuckled. And there were others as well.

This led to a scramble by Heaton for funding to get a proper look at what El Capitan and other caves on the island might contain. Money for travel, and later to pay for student assistants, was always in short supply. He also teamed up with paleontologist Fred Grady of the Smithsonian Institution. Heaton and his wife Julie took a quick trip to Alaska to collect the first of the bear bones in 1991. To reach that key part of the cave, he had to squeeze his way up the tight vertical spots, in part free climbing, pressing himself against the sides through sheer athletic prowess, and gripping the tiniest ledges and clefts with his hands and feet. In other places he resorted to technical climbing, using ropes for safety. When he reached the high passage, a deep pool of water blocked his way, so he braced his arms and legs against the narrow walls of the passage. In pitch blackness, except for his headlamp, he judiciously moved one arm or leg and then the other and "bridged" his way across the little pond.

It was all worthwhile. Spread before him, just as Allred had said, was a field of marvelously preserved bear bones. The cave's natural chemistry had turned them an unusual reddish color. Heaton knew that the bears had not used braided rope and pitons to scale those chimneys—there must have been another entrance to the passage, one that was now blocked by rubble. By doing some precise surveying, posting Julie Heaton to listen outside, and having Kevin Allred shout on a set schedule inside, they found the ancient entrance and opened it. This allowed them to survey the bones properly and collect them. These were the fossils, both black bears and grizzlies, that turned out to date from 6415 to 12,295 years of age.

Heaton and Grady, following in the tracks of the amateur cavers, began searching for fossils in other caves. This led to the discovery of On Your Knees Cave and the bear bones that dated back more than 30,000 years, to well *before* the last peak of glaciation. But, as Heaton had already explained, the idea of a true glacial refugium—an area that had remained ice-free right through the peak of glaciation—was still controversial.

"So I came back to On Your Knees Cave in 1995," although again it was just a case of hiking in for one day, "and did a little more sampling of sediment to try and get a better feel for what was here." This time he had a bit more help.

"We hauled out ten or twenty bags of sediment to screen—to see what the deposits contained." Were there just a few loose bones, he wondered, or was this a site containing major faunal remains? The 1995 bones were sent away for dating. Just like the dates from the previous year, which exceeded 30,000 years, there was again a big surprise.

One bone they had collected was the ulna, or "forearm" bone, from the flipper of a ringed seal, a species that normally thrives in sea ice. "The first date we got was 17,500, which was right at the peak of the glaciation. But it turned out the lab had screwed up a little," Heaton chuckled. The lab had not really prepared the specimen properly for dating, and a second attempt yielded a different result. "It's really a bit older, more like around 19,000." But that still placed it close to the peak of the glaciation.

Over the next few years, Heaton and Grady found other seal bones that dated to around 14,000 years ago. That also put them pretty far back into the glaciation, when according to the textbooks everything should have been smothered in ice. "So we have stuff on either end of the glacial maximum," as well as the much older pre-glacial bear bones. In the late 1990s, Grady was still coming up to Alaska to put in a few weeks of fieldwork each season with Heaton and the rest of his crew, excavating the Seal Passage. "It'll be interesting to see, as we date more ringed seal, whether they span the whole period, or just the beginning and end."

Now, seals, Heaton knew, would not slither up a mile-long trail to over 400 feet above sea level. Their carcasses—or parts of them—must have been dragged up to the cave by predators. "In fact, some of the ringed seal bones have what distinctly look like bear-size canine bite marks. It's really hard to tell which bear. Grizzlies? Polar bears? I don't know. But it does say that the cave was open and habitable. Bears can't go into a cave that's under 1000 feet of ice."

Until the seal bones were found and dated, the oldest Prince of

Wales bear bones *following* the peak of glaciation had been dated at around 12,300 years old, from El Capitan. Located in a glacier-cut valley sloping down to the sea, that cave might well have only become accessible as a bear den once the ice sheets began to retreat from the outer coast. Bear bones from On Your Knees Cave had been dated to 35,000 years and more, or well *before* the peak of glaciation. So the suspicion of a true glacial refuge on Prince of Wales had not previously been proven. Now, with the seal bones from the cave dating to near the peak of glaciation, Heaton thought it was quite likely. He conceded that there was lots of evidence of glaciation on the north end of the island, and even right on the slopes where we were sitting. But he thought it must have been left by earlier glaciations.

A few geologists, such as Tom Ager of the U.S. Geological Survey, begged to differ. Since no bones had been found that spanned the period between 19,000 and 14,000 years ago, the north end of the island might have been overridden by ice for at least several thousands of years at the very peak of glaciation. But Heaton wasn't digging in his heels and being dogmatic about this. Once the project was wrapped up there remained indeed a gap of about 3000 years for which no fossil bones of any species were retrieved from the cave or from any other on the island. In Ager's reconstructions of the glacial flows through the region, the genuine refugia—places that were likely ice-free throughout the entire glaciation—were islands and portions of the continental shelf considerably farther out from the mainland than Prince of Wales Island, especially adjacent to Dall, Coronation, and Baranof Islands.

Heaton also backed off from one of his other early assumptions over the next couple of years. When the tallying and study of all the animal bones were complete, he saw that bears were not the only animals that could have carried the seal bones up to the cave from the sea. He and Grady had also found the bones of arctic foxes. So it was entirely possible that ringed seals were killed at various times by polar bears down on the ice at sea level, but a few bones might have been scavenged and carried up to the cave by small vulpine carnivores. Although there were bear-size bite marks on some bones, these could have been made during the initial predation.

After our lunchtime chat, Heaton went back to his routine. He fired up the generator and suited up for another hour or two of groveling on his knees while grubbing out bags of sediment. And he pressed me into service to help with one of his essential but tedious chores, the initial washing of what came out of the cave.

All the stuff removed eventually had to be screened to separate out the fine bones and teeth, the true gold that was being mined here. When compiled and analyzed, these finds would give Heaton and Grady a complete picture of the cave's fauna. Even the tiniest specimens, such as bones from the different vole species, were indicators of environmental conditions as they changed over time. But along with lots of broken rock and other debris, the small telltale objects were embedded in wet, lumpy, and very sticky clay from the floor of the cave. The first step was to remove this clay.

As each plastic bag of sediment was collected, a strip of tape coded with letters and numbers was slipped into the bag to identify exactly where it had been located on a grid of the cave site. Only then could it be taken outside and left with the other bags earmarked for washing. Since some 5000 bags came out of the cave over the years, washing was a major task that was done at a level spot in the forest clearing on the edge of a dropoff, where the mess made by dumping muddy waste water down the slope would not be a problem.

I put on rain pants over my gumboots to keep my feet and legs dry, then went to join Heaton's assistants, who were seated in a semicircle on inverted five-gallon plastic buckets. Kei Nozaki, the graduate student, showed me how to take a plastic bag of heavy sediment and empty it into a long, narrow mesh bag that looked like an extremely heavy-duty stocking. It was made from nylon mosquito netting, which could take a lot of abuse without tearing. First I had to make sure the coded label stayed with the sample. Next I had to position a large bucket of water between my legs and dip the mesh bag into it.

Then the real work began. Hunched over the bucket, I dipped the bag up and down in the water until it was thoroughly soaked. I got my hands down around it and kneaded it vigorously to break up the largest lumps. The water turned a murky gray color as the clay dissolved. Once the clinging glob was loosened up, I could reach down into the bag, pull out the largest pieces of ordinary rock, which were of no interest to Heaton, and toss them aside. After more dipping and kneading, I began removing the smaller pebbles and broken stones. Finally, all that was left was a wet mass of loose material at the very bottom of the bag, mainly tiny fragments of bone, teeth, shell, and other organic stuff, plus very fine pebbles. I tied off the bag and set it aside, to carry down to the beach at the end of the day for a final washing and drying.

The clay was tenacious stuff. Each bag took ten or fifteen minutes to process, and the bent-over position was hard on the back. But the work gave me a chance to get to know the assistants, all of whom were receiving small stipends for their summer's effort. Nozaki, the oldest of them, had been here the previous summer as well. He enjoyed camping out, except for the occasional bear encounter. (One time during my stay, he came running back to the tents from the distant outhouse, where a bear had interrupted his morning moment of privacy.) And he liked the work, which could have its lighter side, such as the time he found an unusual-looking bone. It turned out to be what keeps a seal's penis permanently erect.

To pay his way through grad school, Nozaki put in many hours each week working in Heaton's lab during the academic year at the University of South Dakota. His job there was to comb through the finished diggings, separating out, identifying, counting, and recording the thousands of fragments of bones and teeth. Because Nozaki was far from home, Heaton took a fatherly interest in him. Space was allotted on the cave project Web site for him to post personal messages and photos that might be of interest to friends and family back in Japan.

Wade Sirles was a tall, strapping twenty-one-year-old who would soon be entering his junior year at South Dakota, majoring in earth sciences. This was his first summer on the island. He had read about the cave project in the student newspaper and hoped the experience would help him decide his academic future and career plans. He had also put in some time in Heaton's lab, looking through the previous year's diggings with a big magnifying glass. "Tim told us what to look for." Sirles admitted that he didn't know all his bones yet; that was part of the learning experience. "To us, every bone we find is, like wow!"

The youngest of the assistants was seventeen-year-old Janna Carpenter, a slim blond-haired high school student from tiny Thorne Bay, about one hundred miles to the south on Prince of Wales Island. "My dad worked with Jim Baichtal, the geologist," she told me. Baichtal brought her up to the site the previous summer for several weeks as an unpaid volunteer to see whether she enjoyed fieldwork, which she clearly did. A couple of years later she was studying under Heaton in South Dakota.

We processed bag upon bag of gooey sludge, hour after hour, as the sun moved slowly across the sky. Sometimes the assistants put on earphones and listened to music. They took turns donning their warm gear and doing digging stints inside the cave, sometimes alongside Heaton, sometimes alone. It was an unusually quiet day at the site, they assured me, because the archaeology crew was away.

Heaton's early work at On Your Knees Cave had involved only short field trips and a few people, which made hiking up from the beach and brown-bagging a lunch simple enough. But when Jim Dixon brought in his larger archaeology crew and equipment, and planned to dig for entire summer seasons, the situation changed. For one thing, Dixon was older than Heaton and not in the kind of shape to tramp the trail each day. Nor was there space for many tents and other shelters behind the little beach. There were chores that needed doing at the cave site, such as tending the large rainwater catchment system. (Water was needed not only for cooking and personal hygiene, but to wet-screen the archaeological diggings.) And a large combined paleontology and archaeology crew, usually totaling ten to twelve people, needed a full-time cook and associated facilities close to the work site. So the archaeologists had set up their own scattering of personal sleeping tents in the nearby woods. The hub of the combined encampment was a prefab yurt, set well off the ground on a wooden platform, that served as a cook shack, office, and communications center for both crews.

But Heaton still preferred to camp down at the sea. And that's where I had decided to pitch my tent as well. It meant a lot of hiking, but there were advantages. It was right on the water, so I could listen to the murmuring waves and squawking seabirds. The sea breeze kept down the mosquitoes. And it was brighter than up in the woods, especially in the evening, when the sun reflected off the water to the west.

After eating dinner at the cook shack, we loaded the day's bags of damp, partly washed sediment into our backpacks and headed down to the beach, but work was not yet over. The students had to dip the bags repeatedly in sea water to give them a final washing. Then they carried them to a large, walk-in drying tent where shelflike screens were warmed by a kerosene heater. Each bag's contents, still identified by its label, was spread out separately on a screen to dry overnight.

Outside the drying tent, Heaton pored over the dry diggings from the previous day, picking out the larger animal bones and teeth. "Here's a marmot's heel bone," he said, holding up a tiny, irregularly shaped object. "Marmots no longer live on the island. And these—" He pulled out a couple of toothpicklike bones. "—are bird humeri, from an auk or puffin." Each time he came across something especially interesting, he put it into a separate tiny plastic bag, labeled it, and filed it away in a cardboard box. The dried residue would go back to South Dakota for the assistants to pick through over the winter.

Bathing Heaton's face in rosy hue was the warm and lingering light of summer at high latitude. His work spot commanded an idyllic view over the beach and dense inshore kelp beds and across placid Sumner Strait. Out in the strait, a broad channel between

Prince of Wales Island and Kuiu Island to the west, salmon jumped and the occasional fishing boat chugged by. One evening, a humpback whale cavorted and spouted, sending up golden sunbursts of spray.

While screening, Heaton went on with the story of how the cave project evolved. In 1995 he found the first seal bone, showing that On Your Knees Cave might have been accessible right through the peak of the glaciation. That same season brought another find that hinted at early coastal occupation by humans. In a low-elevation cave called Kushtaka, not far from El Capitan, Heaton and fellow caver David Love collected some bone fragments. One turned out to be the tip of a bone spear point. They also found a shard of obsidian, a volcanic glass that they knew was not from the island. Both bespoke human presence, and the spear point was in close association with the skeletons of two black bears, which were around 8500 years old. By implication, the spear point and sliver of obsidian could well be the oldest known archaeological finds from the island to that time. (In the end, the spear point turned out to be "only" about 2900 years old, which was not all that ancient, but that dating took a while.)

Meanwhile, though, it appeared likely that human activity on Prince of Wales dated back to the very early Holocene. Jim Dixon had already been following the cave explorations with great interest and now had a reason to become directly involved. Dixon, at the time director of the Denver Museum of Natural History, teamed up with Heaton that winter on a joint application for funding from the National Science Foundation. They were turned down, probably because such ancient human occupation of these outlying islands was considered highly speculative.

In retrospect, given the relatively modest actual age of the bone spear point, the rejection was probably a blessing. A major dig at Kushtaka might have diverted attention and resources from On Your Knees Cave. But despite the apparent setback, Heaton remained bullish. "I feel that we've only seen the tip of the iceberg so far," he wrote in the *Alaska Caver* that year. He predicted that in the coming years both paleontologists and archaeologists would very carefully excavate the deposits in the Prince of Wales caves. And so they did.

Even if the archaeology was delayed, the paleontology was on a roll. On the strength of the seal and bear bones, Heaton was able to get funding from the National Geographic Society to spend two weeks at On Your Knees Cave the following summer. His wife Julie came along, and so did Fred Grady of the Smithsonian. A Forest Service helicopter flew them and their gear to the beach site where we were talking, and they blazed a proper trail up to the cave. Each day they dug out many bags of sediment and hauled them down to the sea for washing. Early in that intensive fortnight of excavation, Grady found their first artifact, a large stone spear point, which confirmed their hunches. People *had* also been in the cave, possibly to hunt the denning bears.

Then came the last day of the excavation, which, like his first trip to On Your Knees Cave, also fell on July Fourth. They had worked all day and were just wrapping up. The others had already finished and hiked down to the beach, where the helicopter was scheduled to pick them up the next morning. Heaton, the straggler, was still inside the cave, digging what he told himself was going to be his last bag of the day. He was wearing waterproof coveralls over layers of polypropylene and working in a corner of the main chamber about thirty feet in from the entrance, an area that was so tight he could barely stretch out his legs. Above him protruded a rocky shelf, so low that he had to lean awkwardly under it and work at arm's length in an area made wet and slimy by the seepage of water from a natural spring.

The battery-powered headlamp on his helmet cast a stark illumination, but digging in soupy sediments meant working mainly by feel. Suddenly, nudged by his trowel, a long, narrow piece of bone flipped up into view. Heaton gave it a quick once-over by the light of his headlamp. Then, reaching in farther, he found another bone, and soon another. Pay dirt. One of the objects he pulled out of the muck looked like a mandible, or jawbone, including the teeth, but it was so caked up with mud that he couldn't identify it. Still, he knew it might be significant. *Maybe this is* not *my last bag of the day*, he thought and decided to dig just a bit longer. He filled two more bags with sediment and carried them down to the sea.

As he washed these last bags, he quickly recognized what he had found. The mandible was clearly human. One of the other bones was part of a human pelvis, which had been chewed up a bit: Predator or prey? he wondered. The first bone that had popped up was about seven inches long and probably the rib of a mammal. (Later analysis showed that it was most likely from a black bear.) The bone had been ground down on the sides and tapered in such a telltale way that it was certainly an artifact of some kind, possibly a punch for flaking stone to make useful implements. Judging by the age of the bear and seal bones he and Grady had been finding, he knew that the human bones and the artifact could prove to be *very* old, quite possibly the oldest ever found on the Alaska coast.

Heaton was excited, but there was a potential hitch. The bones might spell trouble. Under U.S. law, whenever ancient human remains are found in such a setting the work has to stop and the local Native American authorities must be consulted. The local tribes, the Tlingit and Haida, could by law simply refuse to allow the bones to be studied. For scientists this was no merely hypothetical concern.

That same month, in July 1996, two college students were wandering along an embankment far up the Columbia River near Kennewick, Washington. They spotted a human skull and reported it to police. Because it was so well preserved, at first the local authorities thought it might represent a modern death, possibly even a murder. They called in forensic anthropologist Jim Chatters, who quickly collected a nearly complete skeleton that has become infamous as Kennewick Man.

Chatters' initial report characterized the skull and face as having "Caucasoid" features: a long, narrow skull, receding cheekbones, and a high chin. It certainly did not look to him at all like a modern Indian. But a CAT scan examination revealed a healed wound in the pelvis caused by a stone projectile point that was still lodged there. This made it a real puzzle. Was this an early white settler who had been attacked by Indians but had survived? Or what?

When samples of bone were sent off and dated, the results shocked everyone involved. The age was about 8400 radiocarbon years, or around 9500 on the scale of "corrected" calendar years that the mainstream media were more commonly reporting. Regardless of the scale, Kennewick Man was ancient. If he was of "Caucasian" origins—whatever that meant—all theories of who was living in North America 8000 years ago were thrown into question. So, too, were the prevailing ideas of how, and by what route or routes, people might have migrated here. Was it really possible that European-looking people had reached the New World so long before the Vikings? Had they been among the first peoples to colonize the Americas? If so, how might that affect the moral and legal status of Native Americans as the presumed first inhabitants of the continent? The skeleton became a worldwide news sensation.

Anthropologists wanted to study the remains further, but local Indians objected. Claiming their rights under NAGPRA (the Native American Graves Protection and Repatriation Act), they demanded possession of the bones. As Armand Minthorn, an Umatilla leader from Kennewick said, "We already know what happened 10,000 years ago from our oral histories. Our people have been part of this land from the beginning of time. We do not believe that our people migrated here from another continent, as the scientists do." The Umatilla and several other tribes argued that the dead individual had to be one of their ancestors and sought to prevent any further research on Kennewick Man. Because the bones had been found on federal land, they sued the U.S. government to obtain the skeleton and said they wanted to give it a solemn reburial.

A blue-ribbon group of prominent archaeologists and physical anthropologists countersued for the chance to subject this unique specimen to all the analytical tricks in their technological grab bag. They argued that there could be no prima facie assumption that the remains were "Native American" under the meaning of the law, or that any modern group could claim a direct biological or cultural connection to a person who lived so long ago.

With science on one side and Indian identity on the other, the issue became politicized in the worst possible way. Some "Aryan" and white supremacist groups weighed into the fray, arguing that Kennewick Man in fact undercut Indian claims to special legal, political, or moral status. Nor were anthropologists themselves united. For some, "political correctness" required being demonstrably sympathetic to, and supportive of, the Native American position, almost regardless of where that led. In recent years, it had come to include returning artifacts and bones from museum collections. Even prior to Kennewick, some very ancient human remains had in fact been handed over to Indian authorities and reinterred.

The Kennewick case degenerated into the nastiest and most public mud fight ever to pit Native Americans against scientists. Countless editorials and newspaper articles have appeared on the topic, along with several books. Even seven years later, the whole issue was still working its way through the federal courts.

None of this subsequent historical baggage was known to Tim Heaton that Fourth of July, but he realized that his project was in jeopardy. He had to get a message out to the person in charge of excavations in the region, Forest Service archaeologist Terry Fifield, and his only link was the Forest Service radio. Finding human remains was highly sensitive news. As a ham radio operator himself, he knew what might happen if he simply announced it over the air. "Everyone and his dog listens in on the Forest Service channels," he told me. And to make things worse, the beach campsite was a poor location for transmission. Heaton had to walk with his handheld radio far out onto the rocks along the shore and away from the wooded hillside. But, because it was a holiday, he still had no luck getting through to the Forest Service. After some frustration, though, he reached a police officer in Craig, the main town on the island, and asked him to pass along a discreet message for Terry Fifield: "Important artifacts found. Must come in on helicopter tomorrow."

By the time Fifield arrived, Heaton had laid out the clean bones on a piece of cardboard. Barely suppressing his excitement, he watched Fifield's face. He was astonished. "It was neat and exciting," Heaton recalled, "the prospect that these could be the oldest human remains ever found in Alaska."

The fate of the bones unearthed on Prince of Wales Island was very much in limbo as Heaton and the others packed up and flew out that day in the helicopter. But he knew that if the human bones or artifacts were anything like the age of the seal bones—or even as old as the grizzlies from the last few millennia of the glaciation—the discovery would shake some of the most cherished beliefs of American paleoanthropology.

CHAPTER FOUR

Clovis First

It was a gruesome spectacle. The huge proboscidean head, with its floppy ears, long hoselike trunk, and ivory tusks, had already been hacked off. A massive hunk of the carcass lay exposed on the cold, bare ground. Around the somber slab of meat ranged a few warmly dressed men who slashed deep cuts into the skin. One of them braced his feet against the hillock of flesh and tore back a loose piece of the thick hide. Another helped by pushing against this exposed edge. A third man sliced away underneath the skin with a sharp stone blade, freeing the outer flap from the glistening red pulp below. Finally, the man wielding the stone knife cut into the flesh itself. His hand and wrist sank into the bloody morass almost to the elbow as he carved off a large piece of moist meat.

Scenes like this must have been common in late Ice Age America. A few details would have been different, though. There would have been no video cameras posted close at hand, documenting the process. The event would not have made the evening news. Nor would America's ancient hunters have come armed with doctorates in anthropology.

The beast being butchered was no woolly mammoth but the best available proxy, a modern twenty-three-year-old female African elephant named Ginsberg that had died at Boston's Franklin Park Zoo. The humans were clad in old jeans and scuzzy winter jackets, rather than furs. And the point of the exercise was not to procure a supply of elephant steaks, but to learn how hunters might have used advanced Stone Age tools to dismember and process the largest species in their bestiary. When I heard about it, I couldn't help suspecting that, along with its serious purpose, the whole thing had also been something of a lark.

The scientists conducting the experiment included two of America's leading archaeologists, Dennis Stanford of the Smithsonian Institution and Robson Bonnichsen, who was then at the University of Maine but would later found the Center for the Study of the First Americans at Oregon State University. Representing Canada's Museum of Civilization was Richard Morlan, whose excavations in the Yukon Territory had unearthed mammoth bones that he thought might have been used as raw material for toolmaking.

More than a decade later, Morlan still bubbled with pleasure at the memory of this project. "We wanted to see how easy or hard it is to break elephant bones," he told me, "and learn more about how they fracture and flake. That is, how can a skilled flint knapper manipulate this particular mass of raw material." Most stone tools were made by striking a larger piece of stone, such as flint, to detach flakes with sharp edges. Morlan and his colleagues thought that similar flaking techniques might have been applied by ancient people, but utilizing very large bones instead of stone.

So they butchered the elephant, starting with stone tools. "Then, once we got down to the bones, we began working with them and we made new tools out of bone to see how *they* would perform." The scientists also noted the way these fresh, or "green," bones fractured, and compared the behavior of green bone with that of dry, seasoned elephant bones. Those had been sent to them from Kenya by the famous paleoanthropologist Richard Leakey.

What they learned was that the flakes they detached from larger pieces of bone "are indeed useful as butchering implements. You can cut meat with them. They're sharp enough to do that. You can also use them as skinning knives." To Morlan this meant that "people with a small kit of stone tools would be able to bring down an animal such as a mammoth and then enlarge their tool kit by using the bones of the very animal they had killed and were butchering." So, employing both stone and bone tools would contribute to a hunting band's mobility. Once butchering at a site was completed, the aboriginal hunters could abandon the bone artifacts, instead of having to carry the entire tool kit to the next killing ground.

The "Ginsberg experiment" fit nicely with an already well-entrenched image of the First Americans as agile, fast-moving folk who specialized in bringing down mammoths and bison. It was a drastically simplified view of life in the late Ice Age. A few critics countered that ancient people most likely also ate smaller animals and gathered a broad variety of plants as well. But the focus on pursuing large, extinct beasts dovetailed perfectly with our vision of America in glacial times, and it had a sexy, heroic quality that captured the public imagination. Mammoths were to the Pleistocene what dinosaurs were to the Jurassic. Years later, Bonnichsen would title his research center newsletter *The Mammoth Trumpet.*

Still, hunting mammoths and other big game was only the most visually exciting element in the first-peopling scenario that dominated school textbooks for fifty years. There were two other key features. First, the early hunters had supposedly crossed a broad, dry land bridge at the Bering Strait about 12,000 years ago. Second, they had then moved south through an ice-free corridor east of the Rockies. Once accepted, the scenario seemed almost self-evident, but it was an intellectual web woven from threads of pure conjecture as well as solid evidence. It only emerged in stages and against considerable skepticism.

Most interested scientists in the early twentieth century assumed that Native Americans had first come from Asia via the general region of the Bering Strait. Precisely when, though, was still a wide-open question. They need not have arrived at the peak of glaciation, when an actual land bridge would have existed. Given the skills of modern Eskimos, people might well have been able to cross the strait in rafts, skin boats, or other watercraft at almost any time. Maybe they simply walked across the ice in winter. Since no solid archaeological evidence of truly ancient people had been found, many scholars and theorists believed that the ancestors of modern Indians had only come to the New World well *after* the Ice Age.

The most influential authority was physical anthropologist Ales Hrdlicka of the Smithsonian Institution. In Europe, he argued, the bones of some Ice Age people, such as Neanderthals, *looked* quite distinct from those of modern people, much more apelike and primitive. Their antiquity seemed unmistakable. But the American situation was different. Hrdlicka measured the oldest known skeletons in the Americas, compared his data with measurements taken from contemporary American Indians, and found no significant difference. Since there had been no apparent evolution, he argued that the First Americans must have arrived in relatively recent times. His personal estimate was that people had been here for only around 4000 years, and that there was simply no evidence for Ice Age man in the New World. But no one was sure, and the issue was not settled for many years.

In 1923 Jesse Figgins, director of the Colorado Museum of Natural History, received a letter from a Texan named Nelson Vaughn, who described a deposit of large bones that were eroding out of an embankment along Lone Wolf Creek, not far from his home. Figgins asked Vaughn to ship him some samples. They proved to be from an extinct species of Ice Age bison, an animal that was considerably larger than the modern American bison, the misnamed "buffalo," whose vast phalanxes of thundering meat on the hoof were shot down in the millions across the old West.

Figgins sent off an assistant named Boyes to retrieve the rest of the fossils. Boyes found the nearly complete bovine skeleton lying on its side and protruding in places from the embankment, a remarkable specimen. Following standard techniques for removing such large fossils, he began cutting around and under portions of the bony remains, and in doing so dividing the surrounding matrix in which the skeleton was lodged into blocks of manageable size. The next step was to reinforce these often crumbly blocks by encasing them in burlap soaked in plaster of paris. Once the plaster hardened, wooden crates would be built in place around the protruding and partially protected blocks. Then the blocks could be removed from the embankment, the remaining portion plastered, and the crate could be completed for shipping.

At first, the work went along in routine fashion, and Boyes was not paying great attention. He was whacking away underneath the half-formed crate enclosing the first block with a hammer and chisel to remove the excess matrix. Then, suddenly, he hit something hard. It turned out to be a complete and beautifully shaped stone arrowhead or spear point about two inches long. And it was lying in a telltale position, right between two of the animal's neck bones. Unfortunately, he had smashed it into several fragments, a capital offense in contemporary archaeology. No matter, he blundered on. Next, while removing a segment of the bison that included rib bones, a second arrowhead fell right out of the skeleton before he spotted it. This one he managed to misplace somehow, or allow to be stolen. While Boyes was quarrying the last block, yet a third arrowhead came to light, this one situated directly under a femur, or thighbone.

Boyes, apparently something of a blockhead himself, did not recognize the significance of these finds. But Figgins did. He emphasized in his report that the "artifacts were taken from *beneath* an articulated," which is to say intact, "and fossilized skeleton of an extinct bison." This implied that the arrowheads or spear points were at least as old as the bison bones, and in this case probably of the same age.

Figgins then studied several private collections of arrowheads that had been picked up from the surface near Lone Wolf Creek. None were like those found with the extinct bison either in shape or raw material. The ones from the bison were of gray flint and very thin. The surface ones were made from flint of a different color, were thicker and had distinctive side notches for attaching to an arrow shaft. These differences reinforced his conviction that the surface arrowheads were likely modern, while the ones found beneath the bison were much, much older. Meanwhile, a review by his museum of earlier records showed that a stone point very similar to the ones found under the bison had been unearthed in Logan County, Kansas, back in 1895. It, too, had been in association with an extinct bison. It all seemed pretty clear to Figgins. He had discovered an ancient cultural complex with its own particular style of artifact.

Once word got out, Figgins began to learn about other prior discoveries. In 1925 he heard from two men with a ranch near the town of Folsom, New Mexico. They had a collection of bones that had been found almost two decades earlier by George McJunkin, a former slave and self-educated collector who had worked as a cowboy and discovered them lying exposed along the banks of the Cimmaron River. Again Figgins asked for samples, and again most of the bones proved to be from an extinct bison, so the next year Figgins sent assistants to the Folsom site to recover more bones. Here, too, the largest portion of a similar stone point was dislodged while the men were cutting the fossils out of the matrix.

Later that season, the crew spotted a second broken stone point while chiseling out the hard matrix. This time, though, the rest of what looked like a small spearhead—the pointed tip—remained lodged firmly in place, right against a bison rib bone. It was the ideal situation that archaeologists term *in situ*, an artifact found undisturbed and in its original context. The rib and artifact were successfully cut out as a block and sent back to Figgins' laboratory. The lodged tip of the projectile point proved to be a perfect match for the larger piece. And it was so finely made, Figgins gushed that it represented "a quality of workmanship the writer has rarely seen equaled."

But he knew that his colleagues in the cautious world of archaeology would still need convincing. Fortunately, yet *another* stone point was soon found *in situ*. This time, Figgins dispatched telegrams asking the prestigious eastern museums to send representatives to view the artifacts and their context for themselves.

Several prominent experts converged on New Mexico and brushed away the dust to view the stone point in its indisputable and undisturbed location. They all supported Figgins. Still, not everyone was persuaded, at least not until one of those visitors to the Folsom site, Barnum Brown of the American Museum of Natural History, took the evidence back with him to New York and made a dramatic presentation. Speaking to a crowd of scientists gathered at the New York Academy of Medicine, Brown held up several of the finely shaped stone points and announced: "I hold in my hand the answer to the antiquity of man in America. It is simply a question of interpretation."

The consensus in American anthropology quickly swung over to the view that "glacial man" had indeed existed on this continent. Since geological methods of dating in Europe already showed that the last glaciation ended about 10,000 years ago, it was clear that people had been in the Americas far longer than Hrdlicka's guesstimate of 4000 years.

The style of stone point found at the Texas, New Mexico, and Kansas sites came to be known as "Folsom points," and they continued to turn up across large portions of the American Southwest and West. Then came a breakthrough at Clovis, New Mexico, in 1932, where stone points of a somewhat different style, and much larger, were found. Once again, they were initially associated with extinct bison fossils. As at Folsom, diggers found some of them firmly embedded between the ribs of bison.

Archaeologists pored over the Clovis and Folsom points and analyzed the techniques involved by making similar artifacts themselves. They realized that both represented very fine, late Stone Age toolmaking, in which the flints were shaped by careful percussion flaking. The points were much more sophisticated than the crude hand axes of the early Stone Age, or the somewhat more refined heavy flake tools used by Neanderthals in Europe. These American projectile points were extremely thin and easy to break, as the digging crews discovered. They had been made by knocking off an initial large flake and then shaping it further by successively removing long, tapered flakes from each side in a highly controlled way. The flint knappers had used a punch, probably made of bone, and tapped it carefully with a hammerstone or similar implement.

Many patterns of spear- and arrowheads were known to archaeologists, but both Folsom and Clovis points had a distinctive feature.After the initial shaping, additional flakes were knocked off each side by blows aimed at the base of the point. This created grooves that ran up the middle of each face and made it easier to set the point firmly into a slotted wood or bone shaft. The grooves were called "flutes," akin to the concave channels on columns in classical architecture.

Although similar in their fluting, the two types of points were different in an important way. Folsom points could have been either large arrowheads or the points of small spears, possibly hurled by a spear throwing device, or atlatl. Clovis points, though, were usually five to eight inches long. They were apparently the heavy artillery of Ice Age hunters, the points of quite sizable spears or lances. Eventually a variety of other tools, such as stone scrapers, knives, hammerstones, and spokeshaves were also found associated with one or the other type of point. When the hunting was done, meat and plants had to be cut and animal hides scraped clean. But the fluted projectile points remained the most impressive and distinctive of the tools.

Since both artifact assemblages were associated with extinct species, they had to be old. How their ages compared remained unanswered, though, right through the 1930s and the war years, when work at the Clovis site was suspended. Meanwhile, Clovis points kept appearing in other sites all across the American Southwest and eventually turning up in places as far apart as New Hampshire and California. Even more than Folsom, it was a very widely spread tool tradition.

After the war, work resumed at the original Clovis location, and the overall stratification of the site revealed an important pattern. A number of smaller, Folsom-style points were found lying in higher stratigraphic levels than the Clovis ones. In the lower levels were

Clovis points, and these were found in close association with mammoth skeletons. Both types of points were associated with extinct Ice Age mammals, but the stratigraphy meant that Clovis had to be older than Folsom.

The *relative* dating was clear, but not the age of these two tool traditions in *absolute* years. Nor was it possible to compare the age of an archaeological find with the date of events in other places. What correlation might there be, for example, between the appearance of Clovis in America and the existence of a dry land bridge at the Bering Strait? Or between Clovis and the separation of the great northern ice sheets along the eastern flank of the Rockies? The ability to establish absolute dates for artifacts and to compare the dates of far-flung events had to await the revolutionary technique of radiocarbon dating.

Until the early 1950s, determining the age of an ancient object was almost as much an art as a science. Usually, the scientist first had to figure out the age of the stratum, the layer of rock or soil or mud in which the object had been found. For really old stuff, in the tens of thousands or millions of years, this could only be done by estimating how long it would take for various strata of sedimentary rock to form. For relatively recent times, back to the last glaciation, there were two other practical methods.

One was dendrochronology, or tree ring dating. This required counting the tree rings in pieces of old or fossilized wood found buried, for example, along with a stone artifact. For each region of the world, scientists determined telltale sequences of seasonal growth, which were influenced by the local weather each year. A distinctive sequence of colder or warmer years, wetter or drier growing seasons, could be recognized in the growth rings of the old or fossil wood. Some tree species, such as bristlecone pine in the U.S. Southwest, live for several thousand years. Overlapping sequences of rings for ancient, preserved pieces of wood were used to extend the technique right back to the end of the Ice Age.

The logic was similar with varve dating, a technique pioneered in Scandinavia and later applied to North America as well. Scientists would count the telltale annual rings in lakebed clays that were deposited by the seasonal melting of glacial ice. Differential degrees of melting, like tree ring growth, left a distinctive record of warmer and colder years. Using this method, geologists calculated that the last glaciation in Europe ended 10,000 to 12,000 years ago.

These venerable dating techniques were still being used in the early 1960s, as I learned when I took several undergraduate anthropology courses at the University of Pennsylvania in Philadelphia. The facilities at Penn included the fine University Museum, which had one of the strongest archaeology field programs in the United States. At the time, over a dozen major excavations were under way on five continents.

My most exciting course was an introduction to archaeology taught by the museum's director himself, Froelich Rainey, a handsome, square-jawed man with thick dark hair and a doctorate from Yale. He had grown up with cowboys on a ranch in Montana, but in best Ivy League fashion he managed to look at ease even in a tweedy three-piece suit. Rainey had paid his dues in fieldwork, hunting whales with the Alaska Eskimos in the Bering Strait and traveling with them on dogsleds in the dead of winter. Once, to survive a three-day storm, he even had to curl up in a snow burrow with an Eskimo family, whose undiapered toddler peed and shat all over him.

Rainey was arguably the most famous archaeologist in America, because in 1951 he conceived a TV show called *What in the World*, and served as its moderator for fourteen years. Seen by millions every week on the CBS network, the format was to take curious objects from museum collections and have a panel of experts try to identify them. Because the studio audience knew the answer in advance, it could be funny as well as educational. They cracked up when a prominent anthropologist identified a medieval Madonna and child as a woman with a baby that was suffering from adenoids. Another time, an art museum director was sure he was looking at a colonial milking stool from New England, when it was actually used by Eskimos while out hunting seals. Soon the BBC launched a similar program called *Animal, Vegetable, or Mineral*, and the two shows exchanged museum objects in a sort of transatlantic challenge. Once Rainey received a cryptic telegram saying that the Brits were planning a very special program and asking for some particularly interesting artifacts. Too clever by half, the Americans sent over such obscure objects that the English panel failed to identify *any* of them. Unfortunately, the Queen and Prince Philip were in the studio audience that day. She was not amused.

Among the texts Rainey assigned to us was *Dating the Past*, by Frederick Zeuner, which was already in its second or third edition and had long been the bible of geochronology. Its emphasis was on a number of techniques, including tree ring and varve dating, that had been used for decades. Only one chapter in the latest edition dealt with radiocarbon dating, which was just coming to full fruition at that time. And Rainey had played an important supporting role in its development.

In 1939, physicist Serge Korff of New York University sent up high altitude balloons with instruments to count neutrons high in the atmosphere. He discovered that cosmic rays (high-energy charged particles coming from space) were bombarding the upper atmosphere and creating a secondary effect, a shower of high-energy neutrons. Most of this radiation never makes it down to the earth's surface; it is absorbed through collisions with the oxygen and nitrogen in the air. Korff discovered that this bombardment was turning some of the nitrogen to carbon, and not just any carbon, but a radioactive isotope of carbon—carbon 14—that would prove to be particularly useful.

The most common form (or isotope) of nitrogen is nitrogen 14, which has 7 protons and 7 neutrons. When a high-energy neutron hits such a nitrogen nucleus, it knocks out one of the protons and lodges there itself. This changes the atom to carbon 14, a carbon isotope with 6 protons and 8 neutrons. But carbon 14 (also called radiocarbon) is unstable because of the excess neutrons in its nucleus. Eventually, the carbon 14 nucleus decays: one of its neutrons turns into a proton, and an electron (or beta ray) is emitted, which turns it back into an atom of nitrogen 14.

These decay emissions, which can be detected by Geiger counters, are random events; there's no way to predict when any particular carbon 14 nucleus will decay. But over a sample containing millions or billions of atoms, the randomness cancels out, and the rate of radioactive decay is quite steady. The average time it takes for half of all nuclei in any radioactive sample to decay is known as its "half-life." Take carbon 14, for example; after 5730 years, half of the carbon 14 will have reverted back to nitrogen 14.

A young physical chemist in California named Willard Frank Libby read about Korff's findings and had a brilliant insight. Libby was a specialist in radioactivity and in the use of Geiger counters to measure emissions. He reckoned that since carbon 14 was being created in the stratosphere all the time, there should be an equilibrium between the rate of creation of radiocarbon in the atmosphere and its eventual decay into nitrogen. Hence, there should be just as much carbon 14 in the atmosphere today as there was at any time in the past. Moreover, the prevailing winds, he guessed, would spread it from the stratosphere and disperse it throughout the entire atmosphere in the form of carbon dioxide.

Then Libby's mind took another leap. Life on earth is based on carbon. Green plants take in carbon dioxide during photosynthesis and use it to build cells. The carbon from these cells is also incorporated into animal cells when the animals eat those plants. If the amount of carbon 14 in the atmosphere remains steady over time, a plant or animal that was alive thousands of years ago should have had just as much carbon 14 in its tissues as one that is living today. But *only* while that ancient plant or animal was actually *alive*. Once an organism died, the radiocarbon in its cells would begin to decay. Libby realized that this steady rate of decay, and thus the declining amount of carbon 14 in its cell tissue, could be used to measure the time that had passed since the organism's death.

Given the half-life of carbon 14, this means that after 5730 years one half of the carbon 14 will have decayed to nitrogen 14. After

11,460 years, only one quarter of the carbon 14 will remain, and so on. Because of this predictable rate of decay, radiocarbon can serve as a steadily ticking clock that tells how long it has been since the original plant was alive. But because the remaining radiocarbon dwindles to a vanishingly small (and difficult-to-measure) amount, the method is only accurate back to around 40,000 years.

Libby spent most of World War II working on atomic research. By 1946, though, when he was at the University of Chicago, he was ready to develop his theory into a practical method for dating ancient organic objects. It would take a few years of work and a lot of help from supporting scientists and institutions.

To establish a base of comparison for his Geiger counters, he needed to measure the radioactivity of newly created carbon 14. Methane gas was a good source, but there was a problem; most methane is millions of years old and its radiocarbon would have decayed long ago. Libby was forced to take a bizarre step. He held his nose and used methane collected from a Maryland sewer plant. As one biography of Libby quipped, he had to rely on excretions from the good citizens of Baltimore.

Next he had to calibrate his technique by measuring the radioactivity of carbon 14 from old materials of known age, and that's where Rainey came into the picture. He was both the director of a museum that had a wonderful range of ancient organic objects in its collection and the leader of a small advisory committee of archaeologists and geologists that helped Libby to obtain samples from other institutions as well.

Libby needed quite sizable samples. These had to be burned to produce carbon dioxide, which was collected and reduced to pure carbon. This was then smeared on the inside of a Geiger counter cylinder about the size of a milk bottle, while the entire apparatus was surrounded with lead to shield it from background radiation.

Today's AMS radiocarbon dating only requires minute samples, such as a single small seed, or conifer needle, or splinter of bone. But Libby needed roughly an ounce for each test, and tests were usually repeated to boost confidence in the result, which meant destroying significant chunks of organic material from priceless and irreplaceable collections. With Rainey's help, though, Libby obtained a burnt piece of bread from Pompeii that had been buried in ash by the eruption of Mount Vesuvius in 79 A.D. There was also a piece of wood from an Egyptian funeral boat known to date to around 1800 B.C. To avoid focusing only on the ancient Mediterranean region, he got a wood core from a giant California redwood, whose tree rings showed an age of 2930 years, and from a sequoia tree dating to 1000 A.D. All the tests came in pretty well as predicted.

And for the oldest sample, Rainey himself cut a large piece of cypress wood from a coffin in the tomb of the fourth dynasty Egyptian pharaoh Sneferu that had been part of the Penn museum's collection. Libby's calculations put the age of this wood at 4750 years, plus or minus 250. Based on ancient king lists that had been verified by astronomical observations—the ancient Egyptians kept records of when the star Sirius rose in the sky—that pharaoh was believed to have lived 4650 years ago, plus or minus 75 years. The match was pretty good.

Radiocarbon dating worked. Libby published his initial results in 1949, refined the technique in the early 1950s and geared up at his own Chicago lab to receive and analyze materials from a wide range of sciences. Other radiocarbon labs were soon established at leading institutions. One of the first, after Libby's own, was at Penn's University Museum, but the technique was so new and complex that this lab floundered for the first couple of years. Only in the late 1950s did most archaeologists gain real confidence in the method. Libby won the Nobel Prize in chemistry, and archaeology had a powerful new tool.

Scientists have since discovered that the cosmic radiation striking the earth's atmosphere has not been uniform over time, as Libby had assumed. This means that radiocarbon dates do not coincide exactly with real calendar years and that tree ring dating, where possible, is actually more accurate. Typical Ice Age dates have turned out to be 10 to 15 percent older than what radiocarbon dating tells us. Popular publications and media now often give "corrected" dates, but the correction factors themselves are complex and still evolving. For simplicity, most scientists still talk, write, and publish their data in uncorrected radiocarbon years, and I have done the same in this book.

Among the first materials Libby studied in his lab were samples of charcoal from strata associated with Clovis and Folsom points. The Clovis materials, then and since, all fell into the time period between around 10,800 and 11,200 years ago, while Folsom sites were dated to between about 10,000 and 10,900 years. Clovis had a wide geographic spread, but even after decades of searching, no well dated artifacts or human remains had turned up that were older. To many archaeologists this meant that Clovis likely represented the very earliest people to enter North America. The Clovis First idea took root.

Meanwhile, geologists began dating wood and other organic materials found in glacial moraines, which yielded a rough picture of the advance and retreat over time of the great ice sheets. Carbon from undersea sediment cores gave geologists a record of sea level change and allowed them to calculate that the Bering land bridge was inundated around 12,000 years ago, as the Ice Age waned. For anthropologists piecing together the story of when people first entered the Americas, this looked like a key constraining date. Land-based big game hunters could not have come across from Asia much later than 12,000 years ago, they figured. From Alaska, they had to reach the temperate latitudes of North America, south of the ice, by 11,200 years ago, the date of the oldest Clovis finds. Thus, it was assumed—but not proven—that the ice sheets had melted back far enough to open a migration corridor, probably by 12,000 years ago or shortly thereafter.

Archaeologists began to search for evidence of such a migration during a time period that fit the Clovis First scenario. But first, a surprise.

In the late 1930s American archaeologist Junius Bird found stone projectile points with some similarity to the Clovis and Folsom ones at Fell's Cave in South America, near the Strait of Magellan. They were in close association with the bones of extinct Ice Age ground sloths and horses. (Horses went extinct at the end of the Ice Age and were reintroduced to the Americas by the Spanish.) The roof of the cave had fallen in, blocking access and ensuring that the artifacts could not have been brought to the cave later.

Initially, there was no way of getting an exact date for Fell's Cave, but, by the late 1950s, carbon dating there (and eventually for other sites in the region) showed that people had probably reached southernmost South America by around 10,800 years ago. This was remarkably soon after the supposed first arrival of Clovis in North America, which presented a quandary for Clovis First supporters. Could people really have made it so far that quickly? And how to explain why they had moved south so fast?

Paul Martin, a paleontologist at the University of Arizona, rode to the rescue with a novel argument. Martin had studied the changing late Ice Age environment, expecting it to account for the rapid extinction of the Ice Age megafauna. Dozens of species, including giant beavers, huge ground sloths, and massive armadillolike glyptodonts, had disappeared within only one or two thousand years, yet mammoths and other large species had survived equally drastic earlier environmental changes elsewhere. (In fact, more recently mammoths were found to have survived until only 4000 years ago on isolated Wrangell Island, north of Siberia.)

These New World extinctions did not correlate well with the precise timing and sequence of environmental change, but what other factor could account for the loss of so many large species in such a short time span? Martin pointed his finger of culpability at Man the Hunter. "The global pattern of extinctions of all large land mammals appears to follow Paleolithic man's footsteps," he wrote. It reminded him of better studied extinctions, such as the much more recent ones caused by Polynesians when they reached islands like New Zealand and wiped out the huge flightless moa birds. "The loss of many large herbivores in North and South America, apparently suddenly, between 12,000 and 10,000 years ago is consistent with this pattern."

The beauty of this grand theory was that it purported to explain the rapid peopling not only of North America, but of South America as well. Martin pictured the first Clovis hunters emerging from the ice-free corridor near Edmonton, Alberta, around 11,500 years ago. They would be entering a continent full of vulnerable animals that had never been forced to contend with such a wily, intelligent, and well co-ordinated predator as bands of *homo sapiens.* In his view, humans advanced easily in an orgy of hunting, spreading out in a rapidly moving wave across the landscape, expanding quickly in numbers with each generation, and wiping out the large mammals as they went.

Critics countered that people could never have multiplied rapidly enough to fill the continent in only a few hundred years; Martin replied that they didn't have to. He envisaged them concentrating along an ever-advancing "front" of habitation and slaughter, where they continually encountered available prey. He also made some back-of-the-envelope calculations based on population growth elsewhere and the techniques and kill rate of modern hunting bands. He made assumptions about how densely populated an area would have to be, what mammal resources it could support, and how quickly human generations could double and double again in population. One of the first paleoanthropologists to use computers, Martin modeled the process statistically and argued that it was quite possible for a small band of people to emerge from the ice-free corridor, hunt their way south, and reach southernmost South America in less than one thousand years, and possibly in only about 500.

Adopting the Cold War imagery and jargon of nuclear annihilation that was on people's lips at the time, he called this his "overkill" theory. In later versions it became the "blitzkrieg" model of rapid colonization. Martin's tour de force provided the keystone for the intellectual arch that was the Clovis First theory, and the hunting imagery was gripping. As *Newsweek* magazine put it, "The standard explanation of human arrival in the Americas is a stirring tale with mythic overtones, of fur-clad big-game hunters marching out of the far, frozen north to conquer a New World, an Eden whose immense beasts had never before seen human beings."

To the general public in the 1960s and 1970s, and even to university students, Clovis First was appealing. Romantic and vivid, it lent itself to artists' renderings, which, almost without exception, depicted bands of people, clad in furs, and wielding spears. Often they were shown trekking through an almost barren corridor between massive ice cliffs, hunting mammoths and bison as they went. And so the Clovis First model of land-based, extremely rapid colonization of our hemisphere became a scientific orthodoxy almost as self-evident and unchallenged as the biblical dogma that preceded it.

On the North Pacific coast, though, a few scientists had their doubts.

CHAPTER FIVE

Coastal Network

A FUNNY THING happened when archaeologists went looking for evidence to support Clovis First. They came up skunked, although it took most of them decades to recognize and very belatedly acknowledge the lack of solid underpinnings for one of their profession's most cherished beliefs.

If the assumptions on which Clovis First was based were correct, the archaeologists should have found Clovis-type artifacts, or possibly some obvious forerunner of Clovis, near the Bering Strait and in central or eastern Alaska. Some of these would likely turn up at megafauna kill sites, and they would have to predate the appearance of Clovis in the middle latitudes of North America. There should also be artifacts in the corridor area, or at the very least traces of the plants and animals on which migrating people could have subsisted. From the 1960s on, an intense search was made to find such evidence, but it never succeeded. On the contrary, research increasingly confirmed earlier hints that pointed to a quite different prehistory for Alaska. Here, again, my old professor Froelich Rainey played a pioneering role.

In the late 1930s, fresh out of grad school at Yale, Rainey was appointed the first professor of archaeology at the fledgling University of Alaska in Fairbanks. Besides roughing it among the coastal Eskimos on the Bering Strait, he spent several summers in Athabaskan Indian territory, directing digs and conducting the first extensive archaeological surveys in central Alaska. His experience in both places left him with strong doubts about the idea that the First Americans had migrated overland through Beringia.

One winter, he journeyed from the Bering Sea by dogsled with an Eskimo who had invited Rainey to join him on a 250-mile circuit of the man's trapline. Inland, a blizzard engulfed them in a near total whiteout that cut visibility to a few yards. The Eskimo navigated, hour after hour, by the angle of ripples on the snow made by the prevailing wind. Their only chance of surviving was to locate a reindeer-hunting camp seventy miles ahead. The Eskimo developed frostbite on his face, but after darkness had fallen they heard barking dogs and found their way to safety.

On the next fifty-mile leg of the circuit, they lost the axe they needed to chop up the frozen seal meat they were feeding the dogs. The sled dogs became famished and began to fade. As before, they absolutely had to reach the next populated encampment. Rainey began hallucinating. "Hunger, fear, weariness, and the featureless white desert can turn one into a zombie. . . . It was a very long day. Again, in the dark, howling dogs led us into camp and the snug warmth of people, dogs, a seal oil lamp, a boiling pot of ptarmigan, and dried salmon for the dogs." As he wrote, looking back as an old man, "In those endless hours running, or riding on the runners of the sled, I could not help becoming more and more skeptical of all those armchair theories about the original migration of ancient men across the Bering Strait and down through Alaska to continental America. On a map it looks obvious. On the frozen tundra in winter it looks ridiculous." To him, such an overland trek would have required nothing less than "the invasion of man's most difficult environment." The modern Eskimos themselves barely survived there, even with their "thorough and highly developed native science." He was much more inclined to think that people first reached temperate America by boat.

Rainey's musings were based on more than gut hunches. Nothing he found in his extensive digs on the Bering Strait or in Central Alaska seemed like a credible precursor of the Clovis or Folsom points. Nor did his excavations turn up the remains of Ice Age bison or mammoths. As he recalled in 1992, before the recent positive evidence for coastal migration began to accumulate, "Seven years of searching in the Arctic proved nothing about those hypothetical Bering Strait migrations that are supposed to have peopled America at an unknown time in the past. That theory is still accepted as fact although no real proof has yet turned up in the Bering Sea region. My guess is that in the future some other theory will replace it to explain the origin of the American Indians."

As for the ice-free corridor, Rainey was the first archaeologist to carry out a detailed survey of its full length. When the Japanese invaded the Aleutian Islands in 1942, the U.S. Army launched a crash program to build a supply route to Alaska, 1600 miles from the Peace River in B.C. to the Tanana River in Central Alaska, all in one summer. Rainey was assigned to scout the region while the army engineers built the road, which exposed a strip of open ground through what had been trackless taiga forest. He traveled the route several times, in aircraft, on foot, in jeeps, even 250 miles on horseback. But, as he recalled, "The archaeological survey was a 'bust': I found just one flint arrowhead (lost out of my pocket in a jeep ride), and no evidence of that presumed migration of ancients through what was supposed to be an ice-free corridor during the last glaciation."

Even by the late 1980s, after decades of searching and many false hopes, no such evidence had turned up. No Clovis-type tools or other traces of people were found that could be solidly dated to the expected time period. And not only were there no mammoth kill sites, it became increasingly clear that mammoths had disappeared from most or all of eastern Beringia well before people showed up. (The mammoths continued to live south of the ice at least until some 11,000 years ago.)

The most common late Ice Age artifacts in Alaska, as it turned out, belonged to a tool complex that was entirely different from Clovis. In his trial excavations, Rainey found a few flaked stone points that had some similarities to the fluted spear points being unearthed in the Lower Forty-Eight states. These could not be dated in the

1930s, but Jim Dixon studied them more recently, along with others found in central and eastern Alaska, and concluded that none were clearly older than Clovis.

Some of the most interesting and abundant ancient artifacts, though, turned up in a dig led by Rainey right on the university campus in Fairbanks. These were so-called microblades, which bespoke an advanced stone tool tradition that was distinct from Clovis and, in fact, diametrically opposed in the way it exploited stone as a raw material. As Alaska became more thoroughly surveyed following World War II, it was the microblade, or "Paleoarctic," tradition that characterized nearly all late Ice Age sites there. Microblades would also predominate in most sites older than about 5000 years along the north Pacific coast.

I first saw a microblade when I went to British Columbia's Provincial Museum in Victoria (which has since had "Royal" added to its pedigree) to research an article on shell middens. In-house archaeologist Grant Keddie showed me maps of coastal sites in B.C., and led me into a storeroom with shelves full of artifacts. Keddie handed me a very thin sliver of dark, glassy stone about a quarter inch wide and maybe an inch and a half long, with parallel edges. "Careful," he warned, and sure enough, one edge was so sharp I could easily have cut myself with it. He told me about an archaeologist who had needed surgery. To prove just how fine the edges on these ancient stone tools could be, the scientist asked the surgeon to do the cutting with a microblade. It made such clean incisions, the story went, that they healed much quicker than usual.

Next, Keddie let me inspect a microblade core, the specially prepared chunk of stone from which blades had been struck. It was multisided, slightly tapered, and fit easily into my palm. The wider end, or base, was almost flat, while along the sides were numerous concave indentations where the blades had been removed. The way most microblades were made, he explained, was by using sharp blows to drive a bone or antler punch against the edge of the base. Sometimes, though, just pressing hard with the punch was enough to "pop" off a blade. The force exerted on the punch, and the angle of the pressure applied, had to be just right, as did the type of stone selected. On the B.C. coast it was usually obsidian or basalt.

Once these tiny blades were knocked off the core, they could be assembled into a composite tool or weapon. If the toolmaker wanted to make a knife, for example, he would select a straight piece of bone or antler. Along half of one side he would take a stone tool called a "burin," which had a sharp, chisel-like point, and gouge a groove. Into this slot he would inset several of his microblades in a row, wedging their thicker edges firmly into the little furrow. The sharp, opposite edges protruded and served as the knife's cutting edge, and the ungrooved end of the bone or antler was the handle.

If the toolmaker wanted to make a projectile point for killing a large animal, he ground down the piece of bone or antler to give it a sharp tip. Next he gouged slots along both sides, starting at least a few inches back from the point. Then, as with the knife, he would inset several microblades, which protruded from each side and looked a lot like the feathers on an arrow. Finally, this wicked business end of the projectile would be attached to a longer shaft, usually of wood.

It was a highly sophisticated tool technology. In central Alaska, the early tool assemblages that featured it came to be called the "Denali tradition." Once radiocarbon dating was available, and many such sites were found, they turned out to date back as far as around 10,500 years. A similar tradition, probably its direct precursor, was found by Soviet archaeologists in eastern Siberia. This western Beringian tradition was called Dyuktai, and it went back at least as far as 13,000 years, and quite possibly 15,000.

Microblades were highly economical in their use of scarce, and heavy to transport, fine stone. According to one archaeology textbook, a single pound of stone could yield over forty feet of sharp cutting edge if it was made into microblades, compared to only eight inches of edge if turned into large, single-piece knife blades or projectile points. With microblades, a hunter could carry around a few cores of precious obsidian or basalt and make knives or spear points as they were needed over many months. This might be an important advantage, especially during long winters in the far North, when the ground and streams were frozen or covered in snow, and it would be impossible to find raw materials. The same weight of stone could only be turned into a few Clovis or Folsom projectile points, the typical tools of the Paleo-Indian tradition. Paleoarctic was a fundamentally different way of utilizing stone.

The other yawning gap in the archaeological record that spoke against Clovis First was the *total* absence of evidence that people had been in the "corridor" area (along the Mackenzie River or east of the Rockies) prior to about 10,500 years ago. Paleontologists and biologists could not even find traces of animals, or the plant resources needed for animals to subsist upon, that dated to earlier than around 11,000 years. Finally, the geological evidence that supposedly indicated a possible corridor at the relevant time was shaky at best.

I was frankly appalled to see how much bobbing, weaving, and creative reasoning was indulged in by archaeology's prestigious mainstream to justify continued support for the corridor idea in the face of the negative research findings. All too typical were the arguments mustered by Frederick H. West of the Essex Peabody Museum, the top authority on Beringia.

In the early 1980s, West acknowledged that further research had to be done in the corridor and admitted that "nothing remotely resembling a Clovis site" or traces of an artifact tradition that might have preceded Clovis had been discovered north of the ice sheets. He even conceded that none of the bison, mammoths, and other big game that Clovis hunters might have preyed upon had been found in the corridor proper, which he put down as "a stretch of country meagerly endowed in game species and thus one having small attraction for man the hunter."

West further fleshed out his "grim picture" of prospects for life in the corridor by quoting the leading geologist who had studied the area, Archie Stalker. "Conditions [for human beings] would have been formidable under the best of circumstances," Stalker had written. "Fauna and flora would have had to be sufficiently established to supply the food, clothing and firewood required for the inhospitable conditions then prevailing. Ice barriers might still have blocked sections of the route; there would have been large, frigid glacial lakes and tumultuous rivers to cross—these latter swollen by drainage from mountains, local rainfall and the melting glaciers; there would have been chilling winds blowing from the glaciers to face, and extended periods of intense cold as man slowly worked his way [. . .] south through the narrow part of the corridor, not knowing where he was going or what he had to face. There would have been the difficulties of crossing bogs, deltas, spurs of ice, and barren landscapes left by the retreating glaciers."

When I first read this, I thought West was about to abandon the idea of a corridor, or at least downgrade its probability and look for a viable alternative, such as a coastal route. But no. Instead, West and nearly everyone else in the Clovis First camp was determined that, for all the doubts, a viable corridor simply *had* to have existed. His pompous pontifications and feeble attempt at wit made depressing reading.

"A logical if extreme conclusion from this reasoning," as he put it, "might be that the route could not have been used at all. Yet a more potent logic suggests it *must have been.*" [emphasis added] Since West saw no possibility of another route, he made light of this veritable leap of faith. "If ancestral Clovis, arriving from central Beringia, makes its appearance far to the southeast in sub-Laurentian America by some means other than levitation, one is now confronted with the problem of delineating a route of march." Undaunted by his own arguments, West went on to present a disarmingly upbeat scenario. "Presumably, then, the migration of a single band southward through the Mackenzie Corridor, a distance of perhaps [2500 miles], might, no matter what model of movement is invoked, have been accomplished in very short order."

I might have laughed at such mental gymnastics if West had been alone, or without influence, but he was not. C. Vance Haynes of the University of Arizona was America's leading geochronologist and expert on Clovis at the time. Ignoring all the doubts and uncertainties that geologists had about whether a corridor existed *at all*, Haynes, like West, simply adopted the absolute best-case scenario. "After 15,000 [years ago]," he wrote, "deglaciation was rapid, and there appears to have been little in the way of glacial ice to obstruct the passage of man and megafauna through Canada. . . . So by the time megafauna were becoming extinct in Alaska perhaps 13,000 years ago, Clovis progenitors may have been able to continue finding woolly mammoths southward until about 1,000 years or so later when their descendants had reached the Canadian prairie plains and had encountered new species of megafauna." All that blather, and not a shred of evidence. "Where's the beef?" I wanted to know.

The flimsy reasoning and cavalier treatment of scientific facts by American archaeology's supposed intellectual leaders made an impression on at least one budding archaeologist. Knut Fladmark, a graduate student at the University of Calgary in the 1960s, was familiar with the corridor region and its dubious status as a pre-Clovis migration route. (Much later in his career, Fladmark led a major dig at Charlie Lake cave at the southern end of the corridor in Alberta that revealed artifacts 10,500 years old, the closest anyone has come to finding ancient human traces in the corridor region. This was, of course, post-Clovis.) But what really convinced Fladmark to question the Clovis First assumptions was the discoveries he made during a series of excavations on the Queen Charlotte Islands for his doctoral dissertation.

The Charlottes ("Haida Gwaii" to the native Haida people), which extend 180 miles in a roughly north to south direction, are the most remote major islands on the Northwest Coast. Hecate Strait, some 45 to 110 miles in width, separates them from the British Columbia mainland. The strait is both relatively shallow and fully open to the prevailing southeasterlies, a combination that generates really nasty sea conditions and contributes to the archipelago's isolation.

My first trip out there was with my friends Steve Phillips and Marlene Rice on their rebuilt forty-foot yawl *Galatea*. It was their shakedown cruise, a journey to test the boat and themselves in preparation for a long Pacific voyage that would take them far offshore. It proved to be a worthy challenge. A couple of years later I joined them on a leg of their cruise from Panama to the Galapagos Islands. Nothing we ran into there was anything like the fury of Hecate Strait.

As soon as we ducked out from the lee of Banks Island, near the mainland shore, we plunged into the raging tempest. We reefed down our sails as far as we could and still galloped across the wind, creaming along at hull speed, the fastest that the boat had ever sailed. It was a wild and stomach-churning six-hour ride. The seas crested to about ten feet as we reached the very shallow waters near the Charlottes shore. Then we tossed around for an hour in those foaming waves before we managed to spot a buoy that we absolutely *had* to locate. It marked the beginning of a narrow dredged channel across the treacherous submerged sandbar that blocked access to our destination, Skidegate Inlet.

It was a tired, wet, and shaken threesome that cruised past the main Haida village of Skidegate and docked at the sheltered outport of Queen Charlotte City. The "City" part was a joke. There was one main street and a population of about 1000. Steve and I restored the sails and other gear to a semblance of order and hung our clothes out to dry, while Marlene went off to shop for dinner. She returned wearing a T-shirt that proclaimed, "I Survived Hecate Strait." Even people who took the car ferry from the mainland often arrived seasick. Our tumultuous crossing was all too typical.

The width and wrath of Hecate Strait were also a formidable barrier to migrating animals. Over thousands of years, this left the Charlottes with a number of unique, or endemic, species. Those that evolved in biological isolation included the Dawson caribou, which is now extinct, and the largest black bears in North America. Scientists and environmentalists like to call the archipelago the Galapagos of the North.

Most lower slopes of the rugged island peaks have been heavily logged, especially for the fine Sitka Spruce that helped build so many Allied aircraft in World War II. But with a total population of only around 5000 mainly concentrated in four villages, the islands retain an unspoiled feeling that attracts nature seekers. Kayaking the sheltered waters of the pristine southern Charlottes is a magical experience. Today, when the outdoors adventurers have had enough, they can prowl a few artsy boutiques in Queen Charlotte City or drink an espresso. They can even reach the islands by direct daily jet from Vancouver.

When Knut Fladmark arrived, though, it was still the back of the beyond and *terra incognita* for archaeology. The little digging done there had barely scratched the surface, turning up nothing earlier than late prehistoric Haida artifacts. During Fladmark's first work stint there in 1967 as part of a museum field team, test excavations in a shell midden yielded artifacts just over 3000 years old. Then, leading his own effort in 1969 for his doctoral dissertation, Fladmark surveyed about a dozen shore sites on large northern Graham Island.

One site was particularly intriguing. Skoglund's Landing, next to the main island road some forty-six to fifty-six feet above current sea level, had turned up seemingly very simple stone tools that had been flaked on only one side. At the time, when stone tools were found at coastal sites, they usually included more advanced bifaces, such as projectile points, that were flaked on both sides. The unifacial tools, which were deeply buried under almost ten feet of sand and gravel that contained no artifacts or other human traces, hinted at a potentially early age. A consulting geologist thought the deposits overlying the artifacts represented outwash from the local glaciers that had mantled the islands in the Ice Age. If so, the artifacts dated back at least 25,000 years, to *before* the last glaciation, which would have revolutionized New World prehistory. This was Fladmark's first inkling that *very* early people might have traveled along the remote outer coast of our continent.

This interpretation turned out to be wrong. In fact, the overlying deposits testified to a very complex geological and sea level history. With the melting of the ice sheets, world sea level rose dramatically at the same time as the earth's crust under the Charlottes subsided quite rapidly. The combined effect was that rising local sea level passed current sea level about 9000 years ago. Sea level in the Charlottes reached at least fifty feet higher than it is today and stayed high for several thousand years, which meant that Fladmark had found his mysterious artifacts in what was once a raised beach. When organic materials in the deposits were dated, it turned out that the artifacts were more likely 8000 to 9000 years old. This was not nearly as sensational as a pre-glaciation Pleistocene age, but it still pushed back the date of known occupation on the Charlottes quite dramatically. And it kept Fladmark wondering just when people first reached those remote islands.

According to the textbooks, the mainland ice sheet extended right out onto the edge of the continental shelf. But Fladmark already knew that at least parts of the Charlottes had escaped glaciation. American pollen expert Calvin Heusser had studied cores and other soil samples from many places along the North Pacific coast, and they provided evidence of glacial period plant life on the Charlottes and the offshore islands of southeastern Alaska. In other words, there had been at least some offshore glacial refuges. The findings were fairly recent and were disputed by some other geologists, but not by the grad student Fladmark, who had spent all his lunch money one week to xerox Heusser's book on the high-priced copiers of the 1960s.

Then Fladmark looked at the nonexistent evidence for an ice-free corridor and decided that, even if there had been a corridor, it would have been a narrow route between mile-thick ice sheets that disappeared into the clouds. This misery highway would have been so cold, windswept, and barren, he reckoned, that it could hardly have sustained the plant and animal life needed to feed migrating people.

While still a grad student, Fladmark published his first short article very cautiously proposing what would become his coastal migration theory. "An interesting possibility," he wrote, "is presented by the presence of cultural remains" on the Queen Charlottes that are at least 8000 years old, "and the geological indications of ice free areas in the northeast end of Graham Island." He went on to link this explicitly to the history of changing sea level. "During or just after a glacial maximum," he continued, "decreased sea level may have exposed the floor of Hecate Strait and opened a migration route via an exposed shelf along the eastern end of Dixon Entrance to the Alaskan Islands, and perhaps as far north as the Bering Strait region."

Soon Fladmark had his Ph.D. in hand and an academic appointment at Simon Fraser University on the outskirts of Vancouver. With the boldness that came with professional credentials, he pushed the coastal migration scenario and has remained one of its leading proponents ever since.

The first prong of his argument was to underscore just how little evidence there was for a viable ice-free corridor. Even after many years of searching, he pointed out, scientists had found "no significant signs of animal life" older than 10,000 to 11,000 years in the central and northern portions. Even if a gap between the ice sheets existed in time to account for Clovis, he argued, it would have been a "daunting" place, blasted by arctic gusts and blocked by glacial debris and sterile, iceberg-laden meltwater lakes. Could people really have made it through such a frigid wind tunnel? he asked rhetorically.

By contrast, the coast would have offered a far more moderate and biologically productive climate with a wealth of food resources. Many coastal foods, such as bivalve shellfish, kelp, and other seaweeds, were there for the gathering at almost any time of year. The tidepools and exposed rocks alone were a cornucopia of edible animals and plants: crabs, sea urchins, mussels, large barnacles, and sea lettuce. As the saying on the Northwest Coast still has it, "when the tide goes out, dinner is served." Other prey, such as seals, sea lions, and fishes, required more time and effort, as well as the appropriate technology: harpoon heads and hooks, fiber nets, gorges, and fish traps. But none of these exceeded the capabilities of late Ice Age people.

Looking beyond the Charlottes, Fladmark studied the nautical charts and pinpointed other large islands (such as Kodiak Island in the Gulf of Alaska and Prince of Wales Island just to the north of the Charlottes) that were also likely refuges at the tail end of the glaciation. It was not only large offshore islands, though, that could serve as stepping-stones for people migrating along the coast. Even a rocky headland on the tip of a peninsula jutting out from the ice-bound mainland, or a very small offshore island, would do. And even if heavy pack ice blocked off some of the refuges in winter, these way stations might have been available to migrating people at least during the summer months. "Anybody who was able to reach such a location," said Fladmark, "would have had access to all the riches of the north Pacific."

Just as Paul Martin had modeled the movements of inland big game hunters, Fladmark did this for coastal people, arguing that maritime people could, if they wished, move very fast. If they were paddling skin boats such as those used by Eskimos, he calculated that they could travel the entire stretch from the Aleutian Islands to southern Chile in about four and a half years. Why then was there so little evidence of very early people? Simple, said Fladmark. People moving fast in boats leave few traces.

And then there was the problem posed by changing sea level. Most evidence of late Ice Age coastal habitation would have been inundated by later rises. Because most of the likely ice-free areas were now under water, he reasoned, only an extremely well-focused search using new technologies could hope to succeed. He was not, personally, very optimistic that ancient coastal artifacts or human remains would be found any time soon, nor has he played a prominent role in the recent quest. But he kept drumming away with his argument that, when assessing the contending theories about the origins of the First Americans, these considerations should not simply be overlooked.

At one Oregon symposium on human origins Fladmark presented some quite startling and graphic numbers. Assuming a late Ice Age rise in sea level of one hundred meters, he calculated that North America has lost approximately 860,000 square kilometers of late Pleistocene coastal plain. "That's close to a million square kilometers of very desirable, very habitable range that is now under water around this continent alone. That's a land area approximately equal to one-and-a-half times the size of Texas, to use a term of measurement familiar to Americans." So, there was a very simple reason so few known coastal sites *anywhere* were older than 8000 or 9000 years. "I've said it before, and everybody else in this symposium has said it, but it is not being absorbed. Late Pleistocene and early Holocene rise in sea levels simply drowned or washed away evidence of nearly all earlier coastal adaptation everywhere around the world—including the northern Pacific coast of North America."

Fladmark was not entirely alone in thinking that it would be much simpler for Ice Age people to live and migrate along the coast than inland. William Laughlin of the University of Connecticut, whose most important excavations were on the Aleutian Islands, argued in the 1970s that "the coastal area has been everywhere more rewarding and congenial to human occupancy than the interior." He did not specifically claim that the very earliest people on the coast of Beringia migrated by boat, though, only that they must have favored and exploited the coastal habitat. The southern coast of Beringia would have provided quite adequate resources and living conditions for "sea-oriented" peoples, who Laughlin imagined were the "ancestors of the Aleuts and Eskimos." He envisaged them "moving slowly eastward and southward" along the coast, "unaware that most of their territory would eventually be submerged."

By the mid-1980s, further evidence was accumulating that supported Fladmark's position. Biologists Barry Warner and Rolf Mathewes and geologist John Clague collected samples of ancient sediment from sea cliffs at Cape Ball along an exposed shoreline of Graham Island in the Charlottes that faces out on Hecate Strait. They analyzed pollen and seeds from those sediments, along with traces of willow twigs, moss and even beetle fragments. All showed that the northeastern part of the archipelago had been continuously ice-free, at least in summer, since about 16,000 years ago.

Mathewes, of Simon Fraser University, went further. He described the area during the late Ice Age as having an Eden-like environment rich in wildflowers and berries and well supplied with freshwater streams, a place that also would have attracted nesting waterfowl in profusion. He reconstructed the sequence of changes that must have taken place in the local ecology. First there would have been tundralike open plains covered with sedges, grasses, rushes, and other herbs. Then, as the climate warmed, the landscape became a mosaic of dwarf willows and crowberries, along with the grasses and sedges. Finally, about 12,000 years ago, the first pine trees would have begun to colonize the area, and about a thousand years after that, spruce appeared.

Appropriately, this same part of the archipelago, the northeastern corner, enjoys a central place in Haida mythology. Following a primordial flood, Raven, the traditional trickster and transformer of living beings, was wet, hungry, and bored. Then the waters began to drop, exposing a strip of sand called Rose Spit, which sticks out like the thumb of a giant trying to hitch a ride north to nearby Alaska. Raven flew down and found clams, cockles, crabs, sea urchins, and other seafood, so he gorged himself. Sated, but still lonely, he called out, "Isn't there anybody anywhere here?" He heard an answering squeak that seemed to be coming from a huge clamshell half buried in the sand. Looking closer, he saw that the partially opened shell was full of tiny squirming creatures, all trying to hide from the dark shadow that he cast. So he crooned to them in his most seductive voice, coaxing them out of the shell. He told about the rich home that awaited them, full of whales and seals and fish. With hesitation, they finally emerged. These were the first people, who were to become the Haida.

Around the same time that offshore refuges like the Charlottes began to look more credible, Canadian paleoecologist Glen MacDonald showed that earlier radiocarbon dates in the supposedly ice-free corridor were in error. After studying vegetation in the corridor, he argued that no such ice-free migration route had existed until the very end of the glaciation. It would be some years before the question was settled, but the doubts became stronger.

Also weighing against Clovis First was the fact that, throughout the 1970s and 1980s, evidence kept appearing from archaeological sites in the Americas that *seemed* to predate Clovis. The conservative mainstream of archaeologists challenged the validity of these sites and clung doggedly to the Clovis First scenario. Others, though, considered at least certain of the pre-Clovis sites plausible, and some of this minority faction thought coastal migration was the most likely way that people could have made it past the continental ice sheets at a relatively early date. Meanwhile, geological and biological research gradually revealed additional late Ice Age glacial refuges along the Alaska and B.C. coasts.

Theorizing was one thing. Proof was another. Coastal hopping certainly seemed *possible*, but only archaeology could prove that people had actually lived out in those refuges. And how could artifacts be found if most of the territory worth considering was now under deep water?

CHAPTER SIX

Barnacles and Bones

MARINE GEOLOGIST Heiner Josenhans kicked at a rock with his sneaker. "Here's something," he announced, and bent to pick up a sharp piece of dark gray stone. I hustled over from the edge of the forest to have a closer look as Josenhans turned it over in his palm. So did archaeologists Daryl Fedje and Joanne McSporran. One quick look and Fedje knew that the stone was an artifact. "There are a lot of stone tools on the beach here," he said. "A lot of bifaces, such as stone knives, and pieces like Heiner's got, a nice basalt flake," which might have been used for cutting or scraping. But Fedje took a hard line against random and unrecorded artifact collecting. He had no way of surveying the beach at the moment to pinpoint and record the location of the find. So he asked Josenhans to put his stone flake back down exactly where he had found it.

We were on a gravel beach at Richardson Island in the Queen Charlottes. A quarter of a mile offshore was the *Vector*, the 130-foot-long government research ship that had brought us here for a week-long cruise. The main goal of the trip, led by Josenhans, was to carry out acoustic surveys of the sea bottom and take sediment cores to determine the precise history of sea level in the area. But there would also be a bit of time to try to retrieve artifacts and explore the overall potential for doing underwater archaeology in the archipelago.

It was late in the ship's second working day, and this was my first chance to see one of the many shoreline archaeological sites that

Fedje and McSporran had spent years probing and excavating in Gwaii Haanas, the southernmost part of the Charlottes. The year was 1994. That same summer, only a few hundred miles to the north, Tim Heaton was exploring the Alaska caves and just beginning to find hints that Prince of Wales Island might have been a glacial refuge.

The scientists had spent a long day towing instruments back and forth to get images of the sea bottom, and the *Vector* had anchored for the night in sheltered Darwin Sound. After dinner, we piled into the ship's auxiliary speedboat—a Boston Whaler—to look at what had proved to be a site rich in ancient artifacts. The sun had already set, but this was August at latitude 52. There was plenty of rosy light in the sky beyond the wooded hills of Moresby Island to our west.

Fedje's job with Parks Canada gave him responsibility for any archaeological discoveries on this beach and all others in the southern portion of the Charlottes. And the area was just loaded with traces of prehistory. Until Europeans arrived in the late eighteenth century, the Haida had scores of villages and smaller seasonal food-gathering encampments scattered along the sounds, bays, and inlets. Then epidemics of smallpox and other introduced diseases decimated their population. By the late nineteenth century nearly all the surviving Haida had moved north from Moresby Island to the villages of Skidegate and Masset on Graham Island. Around two thousand Haida still live in those villages today. A slightly larger number of nonnatives also call the Charlottes home.

European contact brought some mining and private settlement to Gwaii Haanas, and a lot of logging, but many of these southern areas were still cloaked in dense stands of virgin rain forest. In other places the second growth rain forest had already come back thick and lush on the steep slopes, a dark and nearly impenetrable tangle of conifers that topped out at rocky peaks 3000 to 4000 feet high. The remoteness, sheltered water, and inspiring vistas made the eastern shores of southern Moresby Island one of the most popular sea kayaking areas in North America. Adding to the attraction were the decaying ancient Haida villages, where visitors could still marvel at the carved heraldic figures on standing totem poles and make out the rotting frames of ancient longhouses amid the waxy green salal bushes, tall sword ferns and low spiky foliage of Oregon grape.

Meanwhile, though, clear-cut logging continued, which led to a bitter confrontation that pitted a coalition of environmentalists and the Haida against local logging contractors, a conflict that was resolved only when the Canadian government anted up millions of dollars to compensate the logging companies. In 1988 Gwaii Haanas (southern Moresby and its smaller neighboring islands) became a national park reserve (the "reserve" designation indicating that there are still unsettled Haida land claims). Since then, local archaeology and nearly everything else in Gwaii Haanas have been administered by Parks Canada and the Council of Haida Nations through a joint body, the Archipelago Management Board.

In came Daryl Fedje, who had done his graduate research in the islands. Parks Canada named him archaeologist for Gwaii Haanas, but wet and blustery winter weather made full-time fieldwork impossible, so for the first few years, he retained his winter office in Calgary and flew out to the islands for extended work stints. Today his base is Victoria, on Vancouver Island.

I had first met Fedje the previous day. Going on forty, he had red curly hair and beard, and was tall and wiry. Probably because of his outdoor working lifestyle, he was a lot stronger than I would have guessed. A year later, this saved me from a possibly disastrous injury.

Joanne McSporran had a round face, closely cropped brown hair, and a cheerful, almost bubbly disposition. Although she, too, had a degree in archaeology, public employment rules forbade Fedje from hiring his wife as an assistant. But she was just as intrigued by the work as he was, and often came along on digs as a volunteer. In winter she also tackled the computerized mapping and other detail work. Her main task during this cruise was to monitor and record the ship's position during the survey work, as indicated by the Global Positioning System. On a later *Vector* cruise, McSporran would be the first to examine the oldest artifact ever dredged from the North Pacific.

As the sky gradually darkened over the luminescent water, we sat to talk on a giant driftwood log. It was a balmy and quiet evening. Close by, bright lights sparkled on the *Vector*, our comfortable floating home. Fedje and McSporran filled me in on Gwaii Haanas archaeology, which they had turned into a family adventure.

Fedje's first major task was a five-year project to survey and inventory the numerous archaeological sites in the new Gwaii Haanas park. He and his assistants found literally hundreds of separate locations with scatterings of artifacts. Such an inventory was essential for the park's scientific and management staff to know what was there, whether it needed protection, and how to manage access to it effectively. Some spots, he explained, "may need protection against people. Campfires could do damage. In some places you may have problems with relic hunters. Other places may be disappearing from erosion." In recent years there had been a great increase in traffic to certain parts of the park, especially by kayakers. "The Gwaii Haanas board may decide that visitors shouldn't use particular areas."

While working on the inventory, though, he was also "trying to put some other things into the picture, including coastal migration." This involved collaborating on scientific papers with marine geologists like Josenhans and biologists like Rolf Mathewes. Fedje also wrote information bulletins for the general public on the ancient archaeological sites and more recent Haida heritage sites. The projects brought him into a close working relationship with the Haida, which was rewarding, and a few had become close personal friends. Several of them also collaborated with Fedje as coauthors on published papers; their tribal council funded Haida participation in much of the work; and the artifacts unearthed would eventually reside in a Haida-run museum.

The inventory took many weeks of fieldwork each summer. Much of Gwaii Haanas was fifty miles or more from the nearest settlement, far beyond the last logging road, and reachable only by boat or floatplane. This meant that Fedje was camping out for extended periods with small groups of paid Haida trainees. Often he had extra professional help from fellow Parks Canada staff archaeologist Ian Sumpter.

Sometimes they were joined by Tina Christensen, who was employed as a consultant through a private archaeology firm. Usually they were able to put two or three survey crews into the field at once, in each case teaming up a professional archaeologist with a Haida trainee.

They used Zodiacs to search the hundreds of miles of convoluted shoreline for surface evidence of ancient habitation and food-gathering sites, and carried out countless tests of subsurface soils with shovels or small handheld coring devices. Some of these tests took them up into the forest behind the beaches, and at times there were enough skilled hands to conduct more thorough excavations of selected sites.

Joanne McSporran went along as frequently as possible, and she and her husband often drafted their children as well. Their oldest daughter, Edana, "had no choice," McSporran chuckled. "She started in archaeology when she was four. Her sixth birthday was celebrated at a dig site. Her ninth birthday was also spent camping out on a site." The couple's younger kids, Bryn and Freia, were now four and six, McSporran went on. They, too, enjoyed the hunt for artifacts and had found quite a few valuable items from the distant past. They also pitched in with the tedious job of screening the diggings and sifting through the debris for artifacts or other objects of interest, such as animal bones. The whole family was accustomed to living in tents, cooking on Coleman stoves, and coexisting with the bears and bugs for weeks at a time in the wet and foggy climate.

The next day, back on the ship, Fedje and McSporran pulled out a shoe box to show me some of their finest recent artifacts, including a pristine microblade core. "Edana was the one who found it," said McSporran, beaming with the pride other parents save for announcing that their children have made it into law school. Now seventeen, Edana was so turned on by archaeology, she was spending her summer with a Haida crew on the even less accessible western side of Gwaii Haanas, which fronts on the wild, open North Pacific.

During the summer inventory surveys, Fedje turned up artifacts that expanded his view of local prehistory. Knut Fladmark had found stone tools on the northern part of the Charlottes going back some 8000 years, but Fedje and his teams pushed the dates back by at least another millennium. Just the previous year, they found a stone flake artifact with a barnacle attached on a beach at Matheson Inlet, a deep slash into Moresby Island. Dating showed the barnacle to be 9200 years old. Across Darwin Sound from where we sat, at Echo Harbor, they found a sea otter bone with cut marks on it that made Fedje suspicious. When he examined the bone under the microscope, it was apparent from the pattern of cuts that the otter had been butchered by humans. The bone proved to be almost 9300 years old. Few archaeologists had expected to find people out on such remote islands quite that early. The oldest artifacts on the Aleutian Islands, for example, go back around 8800 years.

The sheer number and concentration of artifacts was impressive. Earlier that summer, Fedje and McSporran had camped out on this very beach for nearly two weeks. All three of their children were on the crew, along with some Haida trainees and a surveyor. They searched for surface artifacts and for buried ones in the woods just behind us. It was an archaeological treasure trove. Many of the gems lay exposed right out on the beach, where the surveyor could map it all with his transit. "We found so many artifacts here," McSporran chortled, "that when they were flagged with red surveyor's tape, it looked like the beach had chicken pox." In total, said Fedje, "we picked up around 2000 artifacts from this beach alone," and all in a stretch of only about 200 yards.

Not only did they find a wealth of artifacts at today's sea level, they were also able to look back in time, at least a short way. The range between high and low tide in the Charlottes can be as much as twenty-two feet. They had picked their work dates so that the most extreme low tides of the year would fall during daylight hours. The gently shelving foreshore meant they could dig test pits well out from the beach and forest and far down in the lowest part of the intertidal zone.

"We had to rush each day as the tide came up," recalled McSporran, "watching the tide pour into the excavations." They knew from the history of sea level rise that the deeper they got, the older the artifacts would be, so they "worked like stink up until the last moment, knee deep in water." And yet, while they continued to find a heavy concentration of stone tools and flakes, the artifacts could not be dated. Not every artifact has a piece of organic matter like a barnacle conveniently attached, so stone tools can usually be dated only by association with wood or charcoal found in the same deposits. Dating is also limited by cost, which can run as high as $1000 a pop. But what they were finding was almost certainly older than the artifacts found higher on the beach. The known history of sea level change meant that the low tide stuff was likely at least 9400 years old. A subsequent major dig in the intertidal zone in 2001 confirmed this date.

The sheer abundance of artifacts seemed to tell its own story. Since artifacts lay nearly everywhere they looked, Fedje did not think he was finding traces of the very first few people to arrive in the area; there must have been time for the first small groups to expand in population. His gut feeling was that people had been there at least a few hundred years earlier, and possibly much earlier than that. It all lent support to the coastal migration scenario.

Anything older than 9400 years would be out there in deeper water. Fedje thought it might be well worth sending down divers to conduct a shallow water excavation. A crude, but simpler, alternative would be to dredge the shallow sea bottom just off this and similar beach sites, which is what he would try doing later on this cruise, using a big set of jaws on the *Vector.* To pursue their quest farther back in time, though, Fedje and McSporran would have to dredge in much deeper waters, and that presented problems. Where they were finding their concentrations of artifacts on shore was in very special places, usually where a stream flowed down to the beach, sometimes fanning out into a small delta. It made sense that people would have camped where they had access both to the beach and to fresh water. But ancient stream and river drainages had meandered all over the place as sea level changed, so Fedje could not just head far out from today's artifact-filled beach and expect to hit an ancient occupation site with the dredging jaws.

To improve the chances of finding significantly older artifacts, Fedje needed accurate undersea maps of where the ancient river deltas and beaches had been 10,000 or more years ago. Without them, dredging for artifacts was like looking for a needle in a haystack. And that's why Josenhans was here with colleagues from the Geological Survey of Canada and their marine surveying equipment. This cruise was just one in a series on which Josenhans would team up with Fedje right through the 1990s. The geology would have to come first, paving the way for the later undersea archaeology. It was a good strategy, and in the end it succeeded, but that's getting ahead of the story.

By the time of our cruise, Fedje already had a pretty good picture of what the ancient environment was like out there under the sea. Studies of pollen taken from sea floor sediment cores pointed to large offshore areas east of the Charlottes with a tundralike environment until almost 12,000 years ago. Then, with the warming climate at the end of the Ice Age, the first trees came in. As they did, though, the sea also began to rise, drowning the lower-lying portions of those newly forested lands. With sea level much lower than it is today, but rising rapidly, the entire shape and character of the shoreline also changed. Lakes that were near sea level were flooded by the sea and became brackish lagoons. The courses of rivers and the locations of beaches themselves shifted, and with them the shorebird colonies, the salmon runs, the places where people could build fish traps. Sea level here rose at least fifty feet in under 300 years, said Fedje. "How many generations is that? Twelve generations? That's not a very long time." He could picture the human consequences clearly. "The places where people lived would not have been at all the same as where their grandparents had lived," Fedje mused. "And each generation would have to pull its canoes higher on the beach."

I enjoyed listening to him paint this vivid picture of ancient life out on these remarkable islands and regretted that on this cruise I would have no chance to see Fedje in action, digging together with the Haida, or to get a sense of what it meant for them to do archaeology on their own territory. Fortunately, a year later, I got the opportunity.

The pilot of our large Turbo-Otter floatplane made several stomach-churning tight passes over the landing area, remote Louscoone Inlet, where a strong wind was kicking up whitecaps. Viewed from above as we circled, they marched like flecks of cotton wool across a narrow ribbon of cobalt blue. Everything else I could see was rugged, uninhabited wilderness. I held on tight as we swooped in low and touched down in a cascade of spray.

It was late summer, and I was back in the Charlottes to spend a week at an archaeology project that was much more emotionally charged than the routine survey and inventory work, which had just wrapped up after five seasons. It involved a crew of fifteen (including cooks and support personnel), who were camped out in tents for two and a half weeks of totem pole preservation. The location was the most isolated of the abandoned Haida villages, Sgan Gwaii. Also known as Ninstints (after an early chief), it has become one of the best known and most photographed native sites in the Americas.

Just reaching the village, which is on a small island near the southernmost tip of the Charlottes, was an adventure. Arriving by jet at Sandspit airport, I had to take a ferry across Skidegate Inlet to Queen Charlotte City and catch the floatplane flight from the harbor. The plane was also taking fresh food and other supplies down to the dig, plus a couple of uniformed Parks Canada wardens who were being assigned to the area for a week or two. On its return, it would rotate out other wardens who had just finished their work stint in the most remote part of the park. Flying south with me was a soft-spoken gray-haired guy with a slight British accent, preservation expert Richard Beauchamp. Formerly with the Royal B.C. Museum in Victoria, Beauchamp had been intimately involved in the physical conservation of Sgan Gwaii and its totem poles for twenty years and was now a semiretired private consultant who raised llamas on his Vancouver Island homestead and during the dig called home on the radio phone to check up on them.

Our flight paralleled the mountainous spine of Moresby Island, a serrated blade of fractured peaks that had been sculpted by glacial ice. Below us was a crazy-quilt maze of narrow fjords, sinuous channels and densely wooded smaller islands. There was no long stretch of sheltered water at Sgan Gwaii itself, so the plane had to set down a few miles away in the nearest inlet, Louscoone, where a Parks boat would pick us up. No safe exchange of people and gear could be made with the propeller turning, so following our wet landing, the pilot had to shut down the engine. By the time the boat reached us, though, the wind had driven us dangerously close to some rocks. The pilot radioed to the boat to keep clear, cranked up the engine again, and taxied us into a position that would allow more time to drift.

Then came a mad race to transfer people, personal belongings, and cartons of food, before the wind again set us down onto a lee shore. The plane had pontoons and struts protruding under the lightly built wing. In a violent chop, these made it difficult to come alongside a twenty-five foot aluminum boat with sharp deck edges and a raised cabin studded with pointed radio aerials and radar. While the wardens passed stuff across the gap, the rest of us fended off as human shock absorbers. I straddled the boat's tossing foredeck, gripping the plane's wing with all my strength to stop it from coming down hard and skewering itself on the boat's electronics. The pilot watched in horror as we drifted ever closer to the rocks, but the transfer was accomplished just in time. One of the wardens fired up the boat's engine. When I mopped my brow and expressed relief, he just shrugged and said this kind of thing was routine here.

We roared out of the inlet, past a promontory studded with wind-tortured trees, and across a short stretch of water fully exposed to the open Pacific. Rolling seas six or eight feet high broke over our bow and surged with fire-hose pressure over our windscreen. On some outlying rocks a half mile or so to the west, great swells broke and cascaded fully fifty or sixty feet into the sky. Their brute force was terrifying. Then our engine raced and sputtered as we crested a wave and the propeller cavitated in the air. I had grim thoughts about what would happen if it crapped out and currents swept us toward those rocks.

It did just that a week later, when I made the same trip in reverse to rendezvous with the plane en route home. I was the only passenger then, and the boat was driven by a Haida park warden named Bev. We had set out from a little bay on the most exposed western side of Sgan Gwaii, which took us uncomfortably close to those rocks. Fortunately, Bev was a trooper. She cursed and sweated and invoked her personal gods, but she also tinkered with the choke and played with the fuel line until she got the engine going again. All the while, I watched the Pacific swells suck back ominously and then explode in paroxysms of foam over jagged pinnacles of stone only a few hundred feet away. They were like a nightmare vision of shark's teeth, lurking to inflict destruction.

The inbound boat ride to the totem pole dig also had a happy ending. After five minutes of jaw-jarring discomfort, we reached sheltered water in the lee of the island, which is a bit over a mile long. The warden who was driving throttled down and eased us into a shallow cove, revealing a dramatic vista. I had seen it in photographs, but it was much more poignant in person.

Dense, moss-bedecked rain forest swept around the crescent-shaped beach like an amphitheater. Looming over a grassy clearing were ragged ranks of carved cedar poles. Some were leaning crazily, others had fallen down. All had been bleached gray and white by more than a century of sun and rain, their sharply hewn lines and shapes softened and rounded by decay. Entire chunks had broken away from many poles, obscuring the figures of eagles and ravens, of beavers and frogs and killer whales.

Once, this had been a major settlement, home to hundreds of people with a bold and vibrant seafaring culture. Once, the poles were painted black and red and white; they stood before great longhouses with gabled roofs and massive beams. Once, huge seagoing war canoes had set out to plunder less aggressive tribes. Then, in the 1860s, the worst in a series of smallpox epidemics ravaged the Haida and the village was gradually abandoned. Now, all that remained of the houses were a few corner posts and rotting rectangles of fallen beams, the weeds and turf growing over them. And yet, the poles still projected a haunting spirit of place. Cracked and crumbling, the carved faces of birds and beasts gazed out at the sea through blank totemic eyes.

Sgan Gwaii, I knew, was unique. For all the decay it was still the best-preserved traditional village on the Northwest Coast. Now an official UNESCO "world heritage site," this isolated village had attracted 1500 visitors the previous summer season, mainly to see the most impressive collection of poles that had not been carted off to museums.

The village had once stood in a clearing, but abandonment allowed the trees to grow back. By 1980, the fast-growing rain forest had engulfed the poles and crumbling house frames, dripping water on them and causing rot to set in. A thick undergrowth of brush had grown up, too, aggravating the problem. Beauchamp, the preservation specialist, studied the site back then and recommended cutting back the forest to let in sun and light and reduce the perpetual dampness. This had been done, and it worked; otherwise, by now there would be no poles left for the project to preserve.

The archaeology crew dropped what they were doing and came down to meet our boat. One of them was Haida tree faller Tom Greene Jr., the guy who had done the actual cutting. Greene was about forty, with dark curly hair, an open, winning grin, and an enthusiasm that had made Daryl Fedje want him on the totem pole project. Like Beauchamp, he was back at Sgan Gwaii again for the first time in ages, and they hugged each other like blood brothers.

As Greene confided to me in a quiet moment a day or two later, the job of falling the trees without damaging the poles and house frames had been very dicey. In the case of one tree, he hesitated and dithered before mustering his courage and tackling the job. It was an especially tricky "snag," a dead standing tree that might break unpredictably when cut down. Greene had to figure out the right way to notch the tree with his chainsaw and drop it precisely in the four- or five-foot space between the remains of two old longhouses. "So, it was kind of scary. I had to get it exactly right." Day after day, he recalled, "I would fire up my saw, and look at the tall snag and shut my saw off and say, *No, I'm not ready for it. It's not going to work.* I didn't have the confidence to do it." At night he fretted about it. Then, one day, the confidence simply came to him and he dropped that tree right in the slot. It was one of the proudest moments of his life.

Coming back now to work on the totem pole dig was something Greene also considered a great privilege. He had not been one of the regular archaeology trainees on the summer survey—mainly younger men and women—so he was delighted when Fedje asked him to be part of the crew. "I mean, how many chances in a lifetime does any person get to be involved in something like this? It's never been done before."

Clearing the forest had helped preserve the poles, but now they were leaning farther each year, and if allowed to fall over they would quickly rot from the moisture and contact with organisms in the soil. The solution was to erect temporary steel cages around each pole, which weighed up to four tons, and dig trenches around the base, which went five or six feet down into the ground. Then, the pole could be pulled upright with hand winches, called "comealongs," and propped up again at the base with rocks and new soil. Beauchamp also wanted to remove small plants that had rooted themselves in nooks and crevices on the poles and were tearing them apart.

Since the ground around the poles was going to be disturbed anyhow, it was a unique chance for archaeological research under Fedje's supervision. On hand to document the work were still photographer Rolf Bettner from the Canadian Museum of Civilization and Jim Taylor, a Parks Canada historian. Capturing it all on video was Barb Wilson, a Haida working for Parks Canada. Even Fedje's boss from Calgary, archaeologist Marty Magne, was there. He liked to get out of the office and dirty his hands with fieldwork.

For more than two weeks, in sunshine and in rain, small teams of diggers with trowels excavated trenches, precisely marked by Day-Glo red strings, around the poles. Down on the beach, others hosed down, screened, and picked through the detritus for anything the diggers might have missed. Day by day, little by little, they rolled back the veil of time, uncovering one telltale artifact after another: iron nails and kettle fragments, pieces of clay tobacco pipes, metal parts of a harmonica, bright blue trade beads, Chinese coins with square holes in the middle, tiny metal thimbles. All bespoke a thriving settlement in the early to mid-nineteenth century, when traditional materials and goods of wood, stone, and bone were giving way to Western ones introduced through trade.

The work was both an opportunity for the Haida to camp out at this special place and a chance for them to embrace archaeology.

"For me," said Tom Greene, "being here is like walking back in time and appreciating how our people lived. Look at the culture, the artwork, the poles with their superbly detailed carving and depth and design." Perhaps because he had begun doing a bit of wood carving himself, just sitting among the poles provided a personal "humility check, a real good one," as well as a deeply spiritual feeling. He looked forward to bringing his wife and children down here someday for a visit.

Barb Wilson, the silver-haired Parks Canada cultural resources manager, also thought the work was uniquely valuable. "Because ours is primarily an oral history," she told me, "the various diseases all but wiped it out." So much was lost, she lamented. "But using archaeology, we can at least look at the artifacts and get a little better idea of, let's say, the economic values that were at play." The poles, she added, "are a reminder of the past and a promise for the future," especially for Haida children and youth.

As during the five-year inventory and survey, which had just ended, this dig teamed up Haida trainees with the nonnative professional archaeologists under Fedje's supervision. Ian Sumpter, the Parks Canada staffer, was big and ruddy-faced, his curly hair held back by a bandana around the forehead. Tina Christensen, young and slim, with long, light-brown hair, was working for the consulting firm Millennia Research. A year or two later she would go back to grad school to get her doctorate in archaeology.

The trainees, even those with no previous experience on the survey work, picked up the excavating technique quickly. "I just have to explain how deep we want to go and where to stop," said Sumpter, dangling his feet into the pit and recording each artifact or piece of bone. "And I emphasize how important it is to have control of the material. You want to be sure you have the three-dimensional context: where you're finding the material—how deep, how far in from the sides—and make sure the information is getting to the recorder."

After a day or two, beginners like Tom Greene and Ernie Gladstone were right into the measured pace of excavation, switching from trowels to fine brushes as objects were revealed. Greene was thrilled to unearth the most aesthetically pleasing find of the dig, a harpoon head probably made of whalebone. Like Greene, Gladstone, whose mainly deskbound job was as manager of services and facilities for the park, had jumped at the chance to be included in the dig. "You start scraping that dirt, and it's almost addicting," he grinned. "You don't know what you're going to find next." The most striking artifact he discovered was the rusted barrel of an ancient flintlock pistol.

Other Haida on the crew were much more experienced. Jordan Yeltatzie was in his late twenties with long, straight black hair tucked into a green baseball cap. He was the nephew of Captain Gold, the Haida elder who first alerted the rest of his people and the B.C. government to the need for action to preserve the poles. Yeltatzie was not much of a scholar. "I dropped out of grade eight," he told me, "and tried logging with my dad. But I only lasted two weeks." He found it to be "real dangerous. I almost got toasted on that job." So he quit and jumped into archaeology, working on a variety of projects every summer since 1985, supplemented by additional months each year in some of the safer forestry jobs, such as tree planting and pruning.

I spent a few hours working alongside Yeltatzie on the beach, screening through the diggings that were being brought down from the trenches. It took a real knack to spot anything of interest in the vast amount of ordinary broken rock and shell. Once, when I was about to dump what was left in my screen, Yeltatzie reached over and pulled out a dark little object that I had overlooked, thinking it was just another rock. It was a human tooth. Yeltatzie set it aside in a special plastic bag for human remains.

Screening could get boring, and what Yeltatzie really enjoyed about the summer surveys was walking the beaches with his eyes peeled, looking for stone tools. "I was really hot back then, looking for lithics on the beach at low tide; I know what I'm looking for," he stated proudly. "Jordan's worth three or four of the rest of us," Fedje confirmed later. "His eyes are just so sharp."

Yeltatzie's big and muscular cousin, Sean Young, had a lot more formal training. His father was in the hereditary line to become chief of one of the long-abandoned villages, Skedans, so a strong interest in Haida heritage ran in the family. Young had put in one summer with Fedje on the survey, and he had a good theoretical background. Interested in archaeology ever since reading illustrated books on ancient Egypt as a child, he was now a third-year student at Malaspina College on Vancouver Island majoring in anthropology and history. He already knew his mammal bones and human anatomy, his stone tools and how they were made. With solid grades in his courses, he was planning to go on to a master's degree in archaeology, and then return to the islands and pursue it as a career.

The high level of Haida participation in archaeology reflected not only their own demands for inclusion, but also an attitude shift among nonnative archaeologists. Working with native groups used to be mainly a response to political pressure, said Marty Magne, whose title was chief of archaeological services in western Canada. "Now it's a sincere commitment." Previously, Magne conceded, artifacts, the best-preserved totem poles from Sgan Gwaii and other decaying villages, even ancestral human remains, were carted off to distant museums, where the Haida had little access to them. Today, all artifacts would come back to the band-run Queen Charlottes Museum in Skidegate after six months of study, radiocarbon dating, and analysis. And human remains were a special case.

Although Fedje directed the day-to-day work, when it came to the remains of their ancestors, the Haida took over. As the poles were straightened and the trenches around each pole were filled again, any human bones that had been found there were given a proper burial. For the first of these ceremonies, Tom Greene took charge. Standing waist-deep in the slab-sided pit, his shirt soiled from digging, Greene lowered his eyes. In his hands was a small wooden box containing a sad scattering of bones. Above him towered a massive decaying mortuary pole carved in the likeness of the mythical sea grizzly. A cavity atop the pole once held the corpse of a high-ranking man. The bones had probably fallen from the pole, or were strewn by animals, to lie for decades until unearthed in the excavation.

Nearby the rest of us gathered. After observing a minute of silence, Greene placed the box in the pit and said a prayer to "our Creator." His voice cracking, he expressed the hope that "You can accept what we are doing here today with the pride and honor that I feel." Tears etched tiny rivulets down the rime of mud on his cheek. The only sounds were the plaintive squawk of gulls and the eternal lapping of the tide.

Then it was back to work. The September weather was good, with only a bit of rain mainly at night, but that might not hold, so there was time pressure to finish the job. Everyone worked hard, but no one harder than Fedje himself. After dinner, when the rest were bagged out in their tents, Fedje would still be at it in the large reading tent, working up his day's notes on the laptop. The Haida admired his knowledge and dedication.

In other ways, though, Fedje had practically gone native and especially enjoyed fishing and boating with his Haida friends. Sunday brought another sunny morning, so Fedje declared it a free day. We could lounge around the camp, flake out on the beach, or go off in the boat fishing or exploring. I decided to join Marty Magne and Barb Wilson to visit what they told me was a wonderful beach on large and uninhabited Kunghit Island. Fedje, Gladstone, and Yeltatzie wanted to try some fishing, so off we went together in the same boat that had rendezvoused with the floatplane. I soon learned how dangerous it could be simply to travel and work in this area.

* * *

Out on the water, the wildlife was spectacular. Seals cavorted by the dozens off some isolated rocks. A pair of peregrine falcons careened across the sky, mating in midair as they plummeted. We could make out the wispy puffs of spray made by some small whales in the distance—Gladstone thought they were minkes. The sea was tranquil and the tide high, and when we got to Gilbert Bay, the avid anglers pulled the boat right in next to some nice dry and conveniently exposed rocks to drop us off. They'd be back in a few hours.

Prowling the long, curving beach, we explored rafts of driftwood logs, some as thick in the butt as I was tall, piled up like gargantuan pickup sticks. Along the water's edge were masses of olive green bull kelp with fat bulbous heads and long whiplike tails. Driven ashore by storms the previous winter, the kelp lay in spaghettilike tangles, each seething with thousands of dancing sand fleas. Along a small river that meandered down to the sea were bear tracks. Fresh ones. The looming rain forest, thick moss, and dense salal made it a scene primeval.

Gradually, the wind picked up. The small waves lapping on the beach built into rollers that thrust with force into the bay. Meanwhile, the tide dropped. The nice little chain of dry boulders where the boat had left us was no longer close to the water's edge. Instead, there were wet, spiky rocks covered in sharp barnacles and slippery seaweed, all swept by each advancing wave. The situation was bad and getting worse when the boat showed up. Magne picked his way out to the most likely exposed rock as the boat jockeyed in close. He leapt and just made it onto the heaving foredeck. Next, Wilson tried, but she slipped and fell. Waist deep in the water, she tore one of her rubber boots on the barnacles and bruised her arm on a rock.

The guys on the boat drifted and assessed the situation while Wilson and I walked halfway around the bay to what looked from a distance like a safer jumping-off rock. But it was no better. Meanwhile, the afternoon was waning fast, the wind was freshening further, and the tide was still dropping, exposing wide swaths of seaweed and treacherous barnacles. There was no option. The boat would simply have to race into the extreme shallows during a lull between waves, and we would have to jump for the bow.

Ernie Gladstone was at the controls, eyeing the waves behind him for the right moment. Fedje was out on the bow, ready to help pull us aboard. "You go first this time," said Wilson. I had sealed my camera and tape recorder in double plastic bags and stashed them in my daypack to keep them dry. As the boat sped in, Fedje signaled to move out onto the farthest boulder. It was alternately exposed and then awash in the seething surf, and covered in slippery eelgrass.

I stepped out onto the rock. First one foot, then the other, and I was poised to leap for the boat. One deft hop and I'd be safe. But then, as I shifted my weight preparing for the jump, I slipped on the seaweed. One gumboot went down, filling instantly with icy water, then the other. First, I was thigh deep in water, then a wave swept in and the water was up to my armpits.

I stared up, appalled, as the boat, several tons of bucking, plunging aluminum, came riding a wave right in on top of me. A nightmare. I was pinned against the rocks by the surging waves, sure the boat's sharp bow was going to crash down and crush me against the boulder, or chop my leg right off. Time stood still. But somehow the boat's first bucking crash missed me. All I could hear—or did I just imagine it?—was the grinding of metal against stone.

"C'mon, c'mon," Fedje shouted furiously. I tried to scramble back up onto my original perch, but the boulder was slick and slimy, and my boots, now filled with water, were like anchors. I flailed around, helpless. Then Fedje reached down and got a firm grip on my wrist. Gladstone put the engine into reverse and gunned it, so the boat began to pull away. And Fedje, in one marvelous burst of adrenaline, pulled me right up out of the water and onto the foredeck. I scrambled quickly past him, crawled through the open windscreen and dove headfirst into the boat. As I lay shivering with cold, the guys maneuvered back in to pluck Wilson off that same damn rock. This time she managed not to slip and came aboard with a lot more grace than I had. I never appreciated the comfort and safety of a tent and warm sleeping bag the way I did that night.

* * *

Back at the poles, the dig continued. The weather remained fair, the meals were hearty, and morale was high. As we heard on the radio each morning, during that very week there were tense armed standoffs between natives and the Mounties elsewhere in Canada. But at Sgan Gwaii, a spirit of mutual respect and genuine comradeship prevailed among the mixed crew.

When the excavation reached bottom, each pole was pulled up straight. The gentle force to do this came from comealongs attached to the corners of the steel cage, tightening up the cables click by gradual click. The guy who directed this part of the action was Tucker Brown, a giant, jovial Haida who operated a large Parks Canada boat. He wore size thirteen boots, a shiny hardhat, and a soft, wistful expression that put everyone at ease. A true jack of all trades, over the years Brown had run heavy logging equipment in the Charlottes and had fished halibut in the Beaufort Sea and tuna as far south as Mexico. Also a skilled welder, he had designed and built the cage-and-comealong system himself, and enjoyed directing the delicate operation of straightening each pole without damaging it.

"O.K. Put a bit more tension on her, Ernie," he called out. "She's swinging a bit. Now slowly. One—two—three—four clicks. O.K., now we've got to go ahead on yours, Jordan. Don't let her get too slack." Gradually the pole, carved in the likeness of an eagle, was ratcheted into a vertical position. Richard Beauchamp checked the alignment with a plumb bob.

To keep the eagle pole upright, the crew hauled wheelbarrows of large stones from the beach. Brown eased them down to Tom Greene in the pit, who wedged them around the base of the pole, where the wood was still remarkably sound. "How's that!" Brown guffawed, pleased at the results. "It'll still be standing long after you and I are ancestors, Tom. After we've both turned to fossils."

When the pit was half-filled with stones, Ernie Gladstone led another solemn ceremony to rebury the human bones found around the pole. The bantering stopped. Everyone was invited to place offerings in the little cedar box with the remains. In went bits of traditional dried salmon, pinches of tobacco, a ring of cedar bark, an apple, even a sandwich. Gladstone added the rusted gun barrel he had found, which might have belonged to the person whose bones were now returning to the earth.

Closing the lid, he climbed down into the pit. Barb Wilson said a few words in Haida. She and Tom Greene hugged and wiped their eyes. The box was passed down to Gladstone, who asked for a minute of silence, then spoke before the box was covered over and the rest of the pit filled in: "I want to thank our spirits for allowing us to disturb them while we do our part in preserving our culture for our children. May they rest in peace here once again."

Nobody talked. People fidgeted and wandered off in quiet contemplation. Standing in the shadow of the poles, I thought about the people who lived here a century and a half ago, whose lives were so different from our own. Then, looking back much deeper in time, I tried to imagine the truly ancient post–Ice Age world. To the east, toward the B.C. mainland under what is now sea, would be dry lowlands covered with sedges and dwarf willows, with mosses and horsetails and crowberry bushes. Then, the first trees would appear, marching gradually across the landscape. And somewhere out there—perhaps—would be a band of people, digging clams and hunting seals along the shore. As the sea encroached, they would follow the food resources slowly, unconsciously westward and upward, generation by generation, until they reached the shores of Gwaii Haanas as we know them today. And each generation would pull its canoes higher on the beach.

CHAPTER SEVEN

Orcas Bring Good Luck

"OK. SOMETHING IS finally coming in," said Heiner Josenhans as he rolled back and forth impatiently on the balls of his feet, staring intently at the long sheet of fanfold paper that clacked its way out of the printer. It was the fourth day of the *Vector*'s research cruise in Queen Charlotte waters, and Josenhans was monitoring a row of instruments set up along a workbench in the shipboard laboratory. Astern of us, the ship was towing a sidescan sonar and an acoustic bottom profiler in narrow Burnaby Strait.

The imaging instruments sent back a real-time record of the shape and structure of the sea bottom. Josenhans was hoping to recognize an ancient depression in the sea floor that had, he suspected, held fresh water for perhaps a thousand years late in the last glaciation. He had obtained images of the spot during a cruise the year before, and now he wanted to take a sediment core—a vertical sample of the sea bottom—to determine precisely when the rising sea had engulfed that location. Dating this event would fill a gap in his record of changing sea level for the area.

"We need a bedrock bump. I think we're coming onto it, Kim," he said, turning to marine technician Kim Conway, my *Vector* cabin mate. "Now, up and over and down into the good stuff," Josenhans urged. The image kept ticking out of the printer, line by shadowy black line, and it looked as though the water was getting deeper. "Come *on!*" said Josenhans, like a gambler throwing dice and coaxing them to come up lucky. "This should be a lake!" But still it didn't look quite right. The lake was a very small target, so Josenhans had to ask the captain, Grant Cadorin, to bring the ship around and try the run again. Several times they tried it. They had the GPS coordinates of the lake and a clear image of the little basin's cross section from the bottom profiler, but it took all morning before they managed to position the ship over the desired spot, 364 feet below us.

Then it was Conway's job to direct the ship's bosun and deck crew, who operated the cranes, A-frames, and winches. They hung the coring device over the ship's side from a massive A-frame, with the top of the "A" cantilevered well out over the water from the edge of the deck. Daryl Fedje, Joanne McSporran, and I came out of the laboratory onto the aft deck to watch. When the coring contraption was in position, it looked like a huge arrow or dart, poised to drop into the water. A long steel pipe pointed downward, like the shaft of the arrow. Attached above it, like tail feathers, was a World War II battleship shell with fins that had been filled with lead. This immense weight would provide the needed downward force. When Conway gave the signal, a deckhand would release the brake on a winch that held hundreds of yards of heavy steel cable, and the coring apparatus would hurtle down through the water to punch the steel tube deep into the seafloor sediments. If all went well, sheer force would fill the tube with a long plug of ooze. The scientists could then bring up a neatly stratified record of the plants, animals, pollen, and windblown dust that had settled into that tiny lake late in the Ice Age.

At least that was the idea, and getting good cores was the most important goal of the cruise. But it didn't work. And it had not worked on any of the previous attempts over the past four days. The top layer of sediment in most places, including here, was just too firm, so the giant arrow only penetrated a little over a foot, fell over on its side, and had to be winched back up to try again. And again. Each time the ship drifted off station and had to be brought back. This frittered away another twenty minutes or more, while Conway raced to set up the next attempt.

Josenhans stood off by himself with a grim look on his face. After the third attempt, he gave up on the target and had the captain move the ship to a different part of Burnaby Strait. "Oh, how I wish I had a vibracore," he muttered to Conway. A vibracore is a larger, more cumbersome coring device that uses an attached motor to help vibrate its way into compacted layers of sand, shell, and gravel. Unfortunately, the available vibracore systems took up too much deck space for the *Vector* and could only be deployed from larger ships. "It's just a little too hard to get down through," Conway agreed. "A vibracore could have done it."

By midafternoon of that fourth day, we were already halfway through the cruise, but Josenhans had not a single good core to show for his efforts. Failure might well threaten his chances of getting scarce ship time for future research. The cost of the ship—salaries for the officers and crew, fuel and other supplies, not even counting pay for the scientific staff—was running at about $5500 per day, and the *Vector* was a relatively small vessel. Other scientists—in marine biology, marine geology, fisheries, climatology, and a host of other specialties—also needed time on the few government ships. Josenhans was lucky if he got three or four days a year, much less an entire week. He paced the deck and fretted. He *had* to bring back some good cores.

The second target for the day was also a freshwater basin that had been inundated by the rising sea at some time late in the Ice Age. Here, too, the exact timing had to be determined by coring, and once again the apparatus failed to penetrate properly. "We didn't hit the feathered edge of that thing," said Conway when the coring device was back up on deck. But it had brought up a bit of mud in the very end of the tube, and they stood at the ship's rail, rubbing it in their fingers. "It's very gritty," Josenhans declared. "Very much like that river sequence we were looking at yesterday, only it's in deeper water. Just picture that landscape, eh, the old flood plain, that flat surface," he said, getting enthused. "It would be really great to date that." The mud, they agreed, was soft enough to be penetrated. "Great," Josenhans decided, "it just might be the ticket." He asked the captain to move the ship just a little so he could try again.

The ship swung into position for a second attempt. The sun was shining, and a stiff breeze kicked up little whitecaps. Suddenly, someone shouted, "Whales!" And sure enough, there they were, the dorsal fins of three orcas, or killer whales, slicing their way through the chop a mile or so away. Josenhans grinned. "Orcas mean good luck," he proclaimed with tongue-in-cheek solemnity. "This core is sure to be a good one. What we need is a little bit of wood in these cores so we can date them."

Lucky the whales proved to be. The second core came up and was hoisted onto the deck. Fedje was the first to reach the lower end of the long pipe and to notice the piece of wood at the bottom. Josenhans had to see for himself. Sure enough, a little chip of wood stuck right out of the mud. "Didn't we just talk about that?" said Josenhans. "Wow!" Fedje scooped out the tiny piece of ancient wood with his fingers and put it into a little Ziplock plastic bag.

"There's gravel in here," said Conway, as he pulled the rest of the core, encased in a clear plastic liner, out of the steel barrel. "Good," said Josenhans, "that's what we want." The core had penetrated about three feet down into the bottom. Josenhans was grinning in relief. "That wood fragment is perfect." At least the cruise would not be a complete write-off. Meanwhile, I was snapping away with my camera. Success at last, which would improve the magazine story I was planning to write. Josenhans turned to me and narrowed his eyes in mock suspicion. "Did *you* put that there?" he laughed.

"I got into this business of marine geology ass-backwards," Josenhans laughed when I first met him. We stood leaning against the rail before the ship sailed. "I was a sailor first, and I still own a boat, a thirty-foot sloop." Now, married with two children, he didn't get out on his boat as often as he would like to. Back then, though, he had left home in Halifax at sixteen and got a job assisting a geologist by ferrying him out to islands off Nova Scotia and helping him pound rock with a hammer. When there was a break in the grunt work, the geologist would explain to Josenhans what he was doing and how the rocks they were chipping away at had been formed. "That really turned me on. I was just thrilled." The geologist found him a job sifting sand in the sedimentary lab at the Bedford Institute of Oceanography, but that was tedious, and Josenhans preferred to be outdoors. "So then I got a job apprenticing to another geologist, learning to map and conducting small-boat surveys of the ocean bottom."

Eventually he did graduate work at Dalhousie University, when geology was booming because of oil exploration and there were plenty of job opportunities. His specialty was mapping, and Josenhans had an uncanny ability to picture the underwater world in three dimensions. "Pull out a profile map of the sea bottom, and I'll bet you a case of beer that I know where it's from." He spent years mapping the geology of the Labrador Shelf and laying out safe routes for undersea oil and gas pipelines. The work had its highs and lows. One time he had to sit freezing for hours in a minisub, desperate for a pee, while the sub operator played Willie Nelson tapes over and over and over. But the sub was tracing deep furrows made in the sea bottom by icebergs, which was exciting. "This business of oceanography gets into your bones," he said. Then he excused himself to get our cruise under way.

Next I met the guy who would be sleeping below me for a week. Kim Conway was a big guy in his thirties with short dark hair, a neatly trimmed beard, and a quiet, almost self-effacing manner. A diver as well as a technician, his dry suit, regulator, and other equipment filled his bunk until he got it all sorted out. And there was marine geologist Vaughn Barrie. Slim and bearded, with a wry grin and trace of a British accent, at the time he lived aboard his sailboat in a Vancouver Island marina. He was coming along initially to help Josenhans, who was based in Nova Scotia but flew out to Vancouver Island every few months. Then, following this cruise, Barrie would take over as chief scientist and lead a cruise of his own, probing the seafloor in Hecate Strait.

Looking after the electronics was technician Ivan Frydecky. And equipped with many underwater cameras, plus his diving suit, was an undersea archaeologist for Parks Canada, Pete Waddell, whose willowy frame towered over everyone else on board. Waddell mainly excavated old shipwrecks in the Great Lakes and on the East Coast, but on this trip he and Conway would assess the potential for an underwater dig in the Charlottes. Finally, there was archaeologist Joanne McSporran. Daryl Fedje, her husband, was already up in the Charlottes and would rendezvous with us there. Also joining us in the Charlottes would be technician Henk Don with his miniature robot submersible.

Around midday we got under way from the ship's home berth near Victoria, cruising north among the myriad wooded islands of the spectacular Inside Passage. To the east, etched against a deep blue sky across the wide Strait of Georgia, loomed the jagged peaks of the Coast Range on British Columbia's mainland shore. Even in midsummer, they were capped with snowfields and glaciers. To the west rose the less lofty but still quite formidable mountains of massive Vancouver Island. By dusk we were threading our way through the narrows north of Campbell River, waters torn by the swiftest tidal currents in the world. That night and the following morning we cruised up narrow Johnstone Strait and out into broad Queen Charlotte Strait, where Vancouver Island and the mainland diverge. Out in the open Pacific, we crossed Queen Charlotte Sound on a northwest course that took us to Gwaii Haanas.

The second day, en route north, Josenhans held a show-and-tell session on the background to the cruise and his research to date. Scientists, the ship's officers, even some of the deck crew crowded into the onboard lab to listen. Marine geology in the area, we learned, had begun with a dramatic find.

In 1987 the *John P. Tully*, a much larger ship than the *Vector*, was taking cores from shallow Cook Bank, ten or fifteen miles north of Vancouver Island, where water depth was 312 feet. The Geological Survey of Canada (GSC) scientists, led by John Luternauer and including our shipmate Vaughn Barrie, hoped at best to find microscopic plant and animal material, such as pollen or diatoms. The goal was to learn what was living in the area at the time the sea floor sediment was deposited. But when they split open the core, they found the hand-size root of a land plant with its wispy ends extending downward embedded halfway down. The *in situ* root was the first concrete proof of what geologists and biologists had long suspected. There were indeed large undersea areas off the B.C. coast that had been dry, ice-free, and quite habitable at the tail end of the last glaciation.

Hitting a sizable and intact plant under more than 300 feet of water was so unusual that one scientific journal quipped that it read like "an accident report"—albeit a very welcome one. Carbon 14 dating showed that the plant had lived at least 10,500 years ago, which meant that Cook Bank was dry land at the time and possibly much earlier as well. As the Ice Age waned, it would have been either a large, low-lying vegetated island, or a major extension of Vancouver Island, depending on where sea level stood at any particular time. And it could have provided ample food resources for animals or people.

In the early 1990s, Josenhans came out to B.C. for a brief West Coast work stint and joined a GSC team that was looking for offshore evidence of ancient earthquakes, but his interest soon expanded to include the search for drowned undersea lands. Over the next few years, sediment cores from a series of research cruises pinpointed a number of such areas off the northern B.C. coast. Between the Queen Charlottes and northern Vancouver Island were Goose Bank and Middle Bank, which were offshore islands at the very end of the Ice Age. East of the southern Queen Charlottes in shallow Hecate Strait was Laskeek Bank, which would have been a major extension of Moresby Island. In fact, 13,000 years ago the land area of the Charlottes was fully twice as large as it is today. And east of Graham Island in the northern Charlottes was Dogfish Bank, which was dry land around 14,000 years ago.

The sea level history of the area was complex and difficult to unravel. During a cruise in the summer of 1992, Josenhans and his colleagues found a low spot in the sea bottom just east of the Charlottes on Laskeek Bank that looked like a shallow basin filled with sediment. One of the acoustic surveying tools, the very sensitive bottom-sub profiler, could distinguish between soft sediments and harder bedrock. In this case, the bedrock of the sediment basin had a depth of 344 feet. Josenhans and his colleagues guessed that the basin might have once been a lake.

Josenhans went over to a large diagram on the wall of the lab and pointed to a profile view of the suspected lake and the adjacent area. "Right off the bat," he said, "we saw a number of terraces, or step-like features relating to when sea levels were lower." (Beach terraces are often formed by wave action when sea level remains stable for an extended period of time.) But the hypothesis had to be tested, so they took a core of the basin. "And that core was just fabulous, a real eureka. It was a full core, ten feet long, filled right to the top." Rather than waiting to take the tube of sediment back to the lab, Kim Conway split the core on the deck of the ship and spotted a little piece of twig about two inches long. This was clear evidence of terrestrial vegetation, and it would provide a large enough sample of carbon to date the layer where it was embedded.

"Then we sent the mud off to Calgary," to have it dated as well. The transition from fresh water to salt water was clearly indicated by the change in diatoms, which are "marvelous, beautiful critters that look like something out of *Star Wars* when seen under a powerful microscope," said Josenhans. "It was the diatom record that confirmed fresh water at the bottom of the lake, then going toward more brackish. Diatoms are very specific to salinity. We also know from diatom analysis that that little twig is associated with the onset of brackish conditions."

At the time, Fedje had his office in Calgary and was doing his master's thesis at the university there. His project involved using diatoms to determine ancient sea levels on shore in the Charlottes. "He heard about our work with this lake on Laskeek Bank," said Josenhans, and that led to their collaboration. Fedje was spending much of each summer on the Charlottes engaged in the five-year inventory pinpointing archaeological sites on land, including above today's beaches. "This was very difficult in the Charlottes," said Josenhans, "because of the thick vegetation. Daryl was totally frustrated. He said, "*We need a rational basis for where to look for sea level in past times.* Until then we're just shooting in the dark."

The problem was a highly unusual geological history in the Charlottes during the late Ice Age, and just after it ended. The weight of the great western ice sheet had deformed the earth's crust, but it did so in such a way that as the ice retreated from the coast, sea level changed in a complex fashion. There had not been a simple rise from the low level at the peak of glaciation to where it stands today. As geologists already knew, for thousands of years sea level in some places had been *higher* than it is today. But exactly how much higher, in which places, and precisely when, were all open questions.

Fedje realized that if sea levels were once higher, he had to look for archaeological sites up behind the beach and back in the dense forest. He was already doing this, but it wasn't easy. The wooded slopes behind the beaches could be extremely steep and the area he was searching was huge. So Fedje and Josenhans decided to coordinate their efforts and create a detailed record of sea level history for the southern part of the Charlottes. They wanted to establish a sea level curve for the area, a graph that plotted the depth of water on one axis, and time on the other. Josenhans went over to another poster on the wall, which showed what they had found so far. The core from Laskeek Bank gave them one point on the curve. At 10,300 years ago, when the little lake basin had changed from fresh water to brackish, relative sea level was about minus 344 feet.

Meanwhile, though, they already had another point for their curve. Fedje and his team had been doing shallow soil probes and test excavations on shore and had been able to date relative sea level in a few spots there. He found that by 8900 years ago the sea had risen well beyond today's shoreline and had reached forty-five to fifty feet *higher* into what is present-day forest. In other words, as sea level rose it passed today's sea level a little over 9000 years ago and kept right on rising. This fit in perfectly with what Fedje and Mc-

Sporran had been finding at current sea level, namely artifacts around 9200 years old. But as Josenhans explained, the sea level story was a bit tricky. After all, once the world's great northern ice sheets had melted, why would sea level continue rising and reach a line that is higher than it is today? The answer was that worldwide (or "eustatic") sea level did *not* keep on rising, or at least rose very little, after around 8500 to 8000 years ago, when the ice sheets had almost entirely melted. But there was more to the phenomenon than a uniform, worldwide sea level rise caused by melting ice.

The earth's uppermost layer, or lithosphere, which includes the crust and continental plates, is somewhat flexible. It floats on the hot and viscous mantle that lies below it. For the northern coast of British Columbia, this means that on the mainland side of Hecate Strait the weight of the ice at the peak of the last glaciation (some 17,000 to 18,000 years ago) depressed the crust. But because the Cordilleran ice sheet never reached out to the Charlottes, on the western side of Hecate Strait the earth's crust was raised significantly in what geologists call a "peripheral bulge" or glacial "forebulge."

"Picture a waterbed," said Josenhans, "and a person sitting down on the edge of it. When you press down here, it comes up over there," where no one is sitting. "That's the so-called peripheral bulge." Now, during the peak of the Ice Age there were in fact glaciers on large parts of the Charlottes as well as on the mainland, but they were local and relatively thin compared to the mile-thick ice sheets that weighed down on the coastal mountains and mainland shore. So the Charlottes area, including Laskeek Bank, was like the side of the waterbed with no one sitting on it. It was uplifted.

Then, as the ice sheets melted and the immense glacial burden retreated from the coast during the time period between around 15,000 and 10,000 years ago, over to the east on the mainland shore the depressed areas gradually came back up. This effect is called "isostatic rebound." At the same time, out to the west, in a mirror image or seesawlike effect, the forebulge gradually settled back down. Meanwhile, worldwide sea level was also rising rapidly from the melting of the ice sheets. The two effects reinforced each other.

The settling of the forebulge took place at the same time that worldwide or "eustatic" sea level rose. This meant that local, or relative, sea level at the Queen Charlottes increased much more rapidly than the overall rate of worldwide sea level rise. It rose well above present-day sea level.

But again, Josenhans emphasized, this was not a case of *worldwide* sea level rising that high. We were dealing here with a localized, flexible, and highly elastic effect. Moreover, the complexity did not end with the settling of the forebulge. Imagine, he suggested, a sloshing wave that continues to ripple through the waterbed when the person sitting on the edge stands up. After the forebulge had settled down for several thousand years, it gradually came back up again—but just a bit—in a sort of final adjustment. This is why after peaking 8900 years ago at forty-five to fifty feet above present-day sea level, and staying there—high in the present-day woods until around 5500 years ago—relative sea level in the Queen Charlottes gradually settled back down to where it is today.

Josenhans continued the show-and-tell session. "So we had gone from minus 344 feet at 10,300 years ago to plus forty-five or fifty feet by 8900, a very, very rapid change in the paleo environment." In fact, so much land on the bottom of gently sloping Hecate Strait (and to a much lesser degree on the steep western side of the Charlottes) was flooded by the rising sea that the entire archipelago shrank to half its earlier size. As one scientific paper described it, the Queen Charlotte archipelago changed from a large land mass of around 1200 square miles "dominated by broad low elevation plains with wide low sloping shorelines," to a much smaller one of around 600 square miles "with a rugged and narrow steep shoreline" where the mountains sloped directly into the ocean.

The entire configuration of the shoreline would have changed many times during this period. Where there were once wide, gently inclined beaches, an area might have changed to one bordered by vertical, clifflike shores. At the same time, new beaches would have formed in other places. The change in narrow, constricted areas, such as valleys where water could be trapped, would have been equally dramatic. Within present-day inlets, places that were once lakes at much lower sea level would have first become lagoons, which are slightly brackish, with the sea flowing into them only at high tide. Then, as sea level continued to rise, they would have become even more brackish estuaries. Finally, they would have turned into fully salt, fjordlike arms of the sea.

Most of the surveying work in the early 1990s had been done from real ships, research vessels like the *Vector* and larger, which could carry the requisite heavy coring equipment. But these ships cost many thousands of dollars a day to operate and were in great demand by government scientists for other projects. Josenhans and Fedje were eager to verify the pattern of sea level change and get more points for their curve, so, for their next couple of cruises, they hired a private fifty-five foot charter sailboat, the *Anvil Cove*, a boat I later came to know well.

She was just big enough to let them do some low-budget surveying and coring in shallow inshore waters. In late 1993 they surveyed Matheson Inlet, a narrow, steep-sided notch at the mouth of a small valley with a bottom contour that was typical of inlets. Carved out by glaciers, in this case a local glacier coming down a valley from higher ground on the mountains of southern Moresby Island, most inlets have a series of submerged but slightly raised sills across them, the vestiges of terminal moraines.

Studying the bottom profile, they found a sill seventy-nine feet deep that ran right across the inlet and would have acted as a low dam when sea level was more than seventy-nine feet lower than at present. On the uphill or landward side of the sill, they reckoned, just before sea level rose to the minus seventy-nine foot stage, fresh water flowing down the valley would have been trapped by the sill to form a small lagoon. Then, the sea would have continued to rise. "This should have meant that the little embayment went from being a lagoon environment to being inundated by the sea," said Josenhans, "and we had to figure out when that happened."

Using quite primitive equipment, they drove core tubes by hand down into the bottom sediment. "We were lucky to get some tree cones and lots of other woody material and shell," all of which allowed various parts of the core to be dated. "The bottom of the core was completely different from the top." It had been deposited earlier and was clearly made up of fresh water sediments. "So we dated the transition from fresh, at the bottom, to marine, when the sea washed over the sill," and that date proved to be somewhere between 9600 and 9700 years ago, which fit perfectly into the picture that was emerging. "Sure enough, that date lay between the lake basin well offshore on Laskeek Bank and the high marine level" that Fedje had found up in the forest. "We got those results in just two days of surveying, which was very lucky." They were doing science on a shoestring.

Meanwhile, working on his archaeological inventory, Fedje had found the flaked stone tool with the barnacle attached that dated to around 9200 years. "And this fits perfectly with the sea level story," exulted Josenhans. "So it's nice to see some human context in association with the change-in-sea-level scenario."

Now they had four points for their curve. At 10,300 years ago, well out in open Hecate Strait, the sea had been very low, 344 feet below where it is today. Then it rose quickly, reaching about seventy-nine feet below present-day sea level 9700 or 9600 years ago and close in to shore. By 9200 years ago it reached today's sea level and kept on rising very fast. And at 8900 years ago it reached forty-five or fifty feet high and far back in the forest, where it remained for thousands of years before gradually settling back to what we know as sea level today.

As Fedje would comment later in the cruise, the sea was just roaring up there, rising at its fastest by as much as several feet each decade.

On the second morning of the trip we reached Lyell Island in Gwaii Haanas and came in close to shore. The tide was out, and the intertidal rocks were encrusted in glistening wet seaweed, a swath of yellow and brown and green. Just above the high tide line was the forest, or at least what was left of it. The island had been heavily logged. In fact, this was one of the areas whose logging had sparked the dispute and protests that led to the creation of the national park. On the hillside above us, old logging roads zigzagged in sharp switchbacks that in some places had eroded, leaving bare soil and rock. The roads led to big clear-cuts. Some were barren, with bleached-out gray stumps that looked like tombstones. Others must have been a decade or more old, because they had begun to green up with new growth of bushes and young trees.

The *Vector* began to run its survey patterns back and forth over preselected areas that Josenhans wanted to map. Conway directed the deckhands in deploying the two main instruments over the stern. Towed behind us, each sent back images to its own monitor and printer in the lab. All the captain had to do was keep the ship on the survey lines, running back and forth and making passes over a different band of seafloor on each run. The scientists called it "mowing the lawn."

The sleek little sidescan sonar looked like a sidewinder air-to-air missile, complete with a shiny cylindrical body and three black tail fins. Flying along like a torpedo at a depth of about 100 to 125 feet, its sound waves illuminated a swath of seafloor whose width depended on water depth. The "towfish" (or simply "fish") sent back a shadowy topographic map of the sea bottom. "There's a trade-off between range and resolution," said Josenhans. The longer the range, the lower the resolution. "We're normally working 325 feet per side, so we illuminate a 650 foot-wide swath. The system is sensitive enough that we can discern the rungs of a ladder lying on the seabed, or a car tire. The sidescan image will help us to locate the paleo shorelines, or deltas or rivers that we will be investigating."

The other instrument was towed along the surface of the water and looked a lot like an inverted stainless steel wheelbarrow bucket set into a ladderlike tubular frame that provided flotation. This was the "subbottom profiler," or simply the "profiler." On its underside was a powerful electrical speaker. "It makes a very discernible click or pulse that sends out a pressure wave," Josenhans explained. "That pressure wave goes down through the water column, bounces off the different layers that make up the subbottom and is then picked up by that stainless steel thing. It's actually a parabolic dish that focuses the energy to a receiver." The electronics discriminate between waves being bounced back from solid surfaces like bedrock and those coming back from soft sediments or empty spaces. This allows it to distinguish bedrock, soft silt, hard surfaces of sizable stones, even domes of methane gas that sometimes permeate the seafloor muds. What comes rolling off the printer as the ship moves along is a longitudinal cross section of the sea bottom.

When both instruments are sending back their images, the scientist monitoring them sees both the overall map of the seafloor from above (the sidescan data) and a cross section of the seafloor immediately below the ship. Learning to interpret the images takes practice, much like reading a radar screen. Over the years, Josenhans told me, he had brought up numerous samples of the sea bottom in sediment cores and with the large "grab" or dredgelike bucket jaws. Then he could check the results against what he had been seeing on the acoustic images. Geologists call this "ground truthing." This gave him a good handle on what to expect, for example, when trying to core a particular area of sea bottom. He could see not only the contours of the bottom, but also what it was made of, how dense it was, and how deep the various layers of sediment were. The firmness of the surface told him how likely it was that the coring apparatus would penetrate. The depth of sediments would show how far it had to punch its way down, and therefore how long a tube to attach to the coring device.

Once our acoustic instruments were deployed, Josenhans went back into the lab and monitored the data coming from the sidescan's printer as the ship ran its survey lines. Vaughn Barrie eyed the profiler printout. The corpulent electronics technician, Ivan Frydecky, watched a video display of the sidescan, which had its own color monitor and was also recording the data on digital tape. And Joanne McSporran was poised with a notebook watching the digital readout on a terminal that displayed the ship's GPS position. Each time the ship turned around while mowing the lawn, she noted the coordinates at the beginning and end of the survey run, as well as at any time that Josenhans asked her to "mark" a specific point of interest. This record would enable the ship to return and target that same spot (to take a core, for example) either on this cruise or in the future.

Surveying looked routine, and at first it was. Then—suddenly—it wasn't.

After several passes by the ship just outside a deep indentation in the shore called Powrivco Bay, Frydecky frowned, threw up his arms in disgust and beckoned to Josenhans. The sidescan had stopped sending back data. Frydecky twiddled the knobs on the video monitor. There were lines and squiggles and static, but no image. Josenhans grimaced. There was nothing he could do but bring the sidescan back up on deck and have Frydecky take a look at it. Meanwhile, the *Vector* continued along the survey line. The bottom profiler, at least, was still working. This part of the survey would just have to be incomplete, but some data was better than none, and there were a lot of other areas that needed to be surveyed during the coming week.

Outside the laboratory, Rob White, the deckhand, winched the sidescan back to the surface. Frydecky and Josenhans swung it up over the side and back onto the deck. Cradling the sleek tube in his arms, Frydecky carried it like a baby into the lab and propped it up against his workbench. Hooking its short cable directly to his video monitor, he worked the keyboard to put the sidescan through a series of diagnostic procedures, looking at one menu after another on his screen and trying to figure out what was wrong. He rubbed his hands up and down along the housings that covered the sidescan's hydrophone receivers and saw that it was, in fact, recording the vibrations. But something was still wrong.

He got out his tools, loosened a few screws and opened up the casing of the towfish. Pulling away the sides and waterproof gaskets, he exposed the top half of the cylinder, with its tightly packed stack of round circuit boards, each one a maze of multicolored components. All Frydecky could do was to check each circuit and, if need be, replace it. Digging under the counter, he pulled out a big cardboard carton, with "Towfish Parts" written on it in magic marker.

Frydecky was a *big* man in every dimension. He hunched over the little torpedo like a shortsighted bear. Squinting into the electronics with his Coke bottle glasses, he pulled out and replaced one circuit board after another. After each change, he booted up the computer program again and gently stroked the hydrophone housings to test the response.

Meanwhile, as the ship kept moving along its survey lines, the first morning of research time ticked away. Josenhans paced the floor of the lab, trying not to look too concerned. He was not happy, but these glitches came with the territory, and he was not really fazed. At least, not yet. I wandered out on deck to get away from the tense atmosphere and enjoy the sun. Then, word came from the lab that the subbottom profiler had also packed it in. I stepped back inside. "It's all buggered up," said Barrie in quiet dismay, although the profiler's printer was still cranking out the fanfold paper.

Frydecky kept beavering away at the workbench, trying to fix the sidescan, while everyone else went for lunch. Meanwhile, Daryl Fedje arrived, coming out from shore in a Parks Canada speedboat to rendezvous with the ship. After a bit of discreet snuggling with McSporran, he unloaded his personal gear and joined the scientific team.

With the instruments on the fritz, Josenhans abandoned the survey work and decided to take the first core of the cruise. "We'll try here in twenty-seven fathoms," he said, pointing to the chart. The target was a spot within the narrow confines of Powrivco Bay. "We did this zigzag pattern," he said, showing me a chart with survey lines on it made during one of the earlier cruises. Then he pulled out the relevant sidescan printouts showing the overview of the sea bottom, and the bottom profiler printout giving a cross section of the seafloor itself.

"The target here is stratified sediments." He pointed to a sill made of bedrock, a small raised hump on the printout that ran across what was once a valley when the sea level was much lower. Even with today's much higher sea level, the bay still had the shape of a valley, with steep slopes on each side and a much gentler slope at the head of the bay, where several large streams flowed down into the sea. In behind this bedrock sill, on the shoreward side, was a place where sediments, eroded down from the slopes above, had filled in. "You can see the old axis of the valley, the drainage system. Here's a piece," he said, pointing to one printout, "and here's another example." He pulled another section of printout onto the chart table for a better look. Each of these spots on the sea bottom was a place where, for perhaps a few hundred years as the sea level rose, the sill running across the valley had created a little, mainly freshwater, lagoon just above sea level. Then, in each case, the lagoon was inundated by the rising sea. The goal was to determine when that inundation took place.

"I think we'll go for the deeper one," Josenhans decided. "It's a broader target. See that little platform just outside the mouth of this bay?" He pointed to a small notch in the shoreline—a bay within the larger bay—where one of the streams came down off the hillside. "There's quite a clear dropoff dip." Josenhans consulted with his geologist colleague, Barrie. After studying the bottom profile, they agreed that there was about a three-foot layer of mud on top of the older sediments. Not too thick for the coring system to penetrate, they hoped.

But they were wrong. The coring device did not penetrate, and this was to be the pattern for the next three days, until those orcas showed up, bringing good fortune at last. Then, amazingly, the coring went smoothly for the rest of the cruise, and Josenhans ended up with excellent data to fill the gaps in his sea level curve.

Meanwhile, though, Frydecky spent all that first day trying to figure out what was wrong with the survey instruments. Even when the ship anchored for the night and shut down its engines, Frydecky was still in the lab, guzzling coffee and hunched over the instruments and repair manuals. Around midnight, he called it quits.

Early the next morning, though, he was back at it, and finally got to the root of the problem. After all the grief—all the replaced circuit boards, all the tinkering with programs and menus—it turned out that there was nothing fundamentally wrong at all. Desperately trying an unlikely solution, he plugged the sidescan into a power cord coming from a different electrical circuit than the one being used by the subbottom profiler, and both worked perfectly. Interference from the profiler, it seems, was causing problems for the sidescan's recorder. All Frydecky and Josenhans had to do was to move the profiler's power cord and printers, so the two sets of instruments were working on separate circuits. By the time the second day's surveying began, everything was up and running just fine. Josenhans had a grin plastered across his face. "How's that for high tech?" he joked. "You move it to the other side of the lab and everything works."

And so the cruise proceeded, with most of the time devoted to routine survey work and coring. In a number of important places, though, Josenhans knew his coring would have to wait until he got time allocated to him on a larger ship with a vibracore. The equipment he had on the *Vector* just couldn't get down into the firmer sediments.

On the third day, technician Henk Don rendezvoused with the ship and brought on board his little robot submersible. The idea was to see how useful it might be in locating possible archaeological targets on the bottom. Later in the cruise, the ship's divers, Conway and Waddell, used their scuba gear a few times to check out the sea bottom in places that Fedje thought might lend themselves to underwater archaeology. Fedje and McSporran were also allotted time in a couple of places to use the "grab" to try to bring up artifacts from the bottom. But that's getting ahead of the story.

For me the most memorable, and rueful, part of the cruise was when I personally screwed up.

One afternoon, while the *Vector* surveyed, Fedje organized a shore expedition that involved the entire scientific team and part of the deck crew. For years he had been trying to locate the exact maximum level the sea reached on shore some 8000 to 9000 years ago. To do so, he had studied a lot of small sediment basins—little lakes and ponds on shore—that looked like they were between sixteen feet and sixty-six feet above sea level.

There was one particular tiny lake on southern Moresby Island that he had never got around to examining closely, and now was his chance. It was on a small peninsula a half mile inland from one of his best shoreside archaeological sites, Arrow Creek. The lowest shore around the lake, as shown on the topographical maps, was about seventy feet above sea level, but these maps were only accurate to around sixteen feet, and Fedje suspected that the elevation of the lake was actually less. "It's just borderline," he explained, and he suspected that the sea might just have flooded into the lake basin as it reached its highest level. "If so, the diatom record will be very clear," said Fedje. "It's a very sensitive record, picking up changes in water chemistry. So that would give us the almost exact sea elevation at its peak, and also a record of environmental conditions—what was growing—as shown by the types of pollen."

He needed to core that little lake, but to do so would take a lot of help. We crammed into the ship's boats with a pile of gear, including two small inflatable rubber rafts, a bunch of wooden two-by-fours, two sheets of plywood, and some long white tubes of four-inch diameter plastic pipe, the kind used for domestic sewer lines. The plan was to carry everything up through the woods to the lake, inflate the rubber boats and assemble a raftlike platform that Fedje could paddle out to the middle of the lake. There he would drive the plastic tubes down into the lake bottom and extract a core. It should have gone smoothly, but it didn't, and I was the goat.

I was assigned the three lengths of sewer pipe to carry up the trail, a light enough burden, I thought. But we had landed well down the beach from the trailhead leading up to the lake. I dawdled, changing from gumboots to running shoes, snapping pictures of the bay, and rearranging the stuff in my daypack. Meanwhile, the others grabbed their loads and followed Fedje, who knew the way. When I picked up the tubes and began to follow, I found they were so long and slippery that it was almost impossible to carry them. The end of one would dip down and dig itself into the sand, while the other end swung skyward. I wondered how I could ever maneuver these ungainly things through the forest.

I had some tape in my daypack and figured it would go easier if I taped their ends together. I was right, but that took more time. When I finally set off in earnest, the rest of the team were far up the long beach and disappearing into the woods, and by the time I got there it was not at all clear just where the trail began. There were several places that had been trampled and looked like the beginnings of trails.

I picked the most obvious one and headed into the woods, carrying my awkward load, and for the first few hundred yards, I was sure I was on the right track. It was definitely a trail, and it was curving uphill onto a bluff. Pretty soon, though, I found it hard to push my long bundle of pipe through the underbrush. I saw no signs of broken branches or recently trampled foliage. Worse, I noticed a few black pancakelike droppings that I took to be bear scat. And I recalled that the archipelago's black bears, which had evolved in isolation in their offshore refuge, were the largest in North America.

I was lost in the woods on a bear trail. The way I was heading, if I didn't run into a bear first, I reckoned the trail would lead me well past the end of the little lake. And then where? Probably across the peninsula and down to tidewater on an isolated inlet where no one was likely to look for me, at least not before dark.

One option was to turn around and head back down to the beach, where I could wait until someone came back to search for me. That would be embarrassing, and would cause quite a delay, because they couldn't take the core without the tubes I was carrying. The other option was simpler, to scream for help, which is what I did.

"Hello! Hooo! Hey! Hey!" I shouted with all my breath. Again and again I called, each time waiting and listening. After five or six tries, I heard something. A voice. Someone shouting back. I called again, and again I heard the faint reply. Heading toward the sound meant bushwhacking through thick undergrowth, but that's what I did. I thrust my battering ram of plastic pipe ahead of me and ducked to keep the branches out of my eyes.

After a few minutes, I stopped and shouted again. Again, a reply, this time noticeably closer. It was a female voice—Joanne McSporran. I crashed on farther until the woods got much thinner. A bright sky! The lake! At last, there she was, sitting in one of the inflatable boats along the shore. "Boy, am I glad to see you," I said. The rest of the team were a quarter mile farther along the embankment, but there was no trail along the lake. So I got into the boat, and she paddled me along to join the others. The group kidded me a bit, but they were so busy assembling the raft that the moment passed. Daryl got out into the middle of the lake and obtained a perfect core. I'd never hear about the embarrassing incident again, I hoped.

Fat chance! Years later, all the key specialists assembled at a seminar in Victoria to review the archaeology work in the Charlottes. I'd been invited, too, and there were scientists I'd never met before. Fedje was showing aerial slides of that little lake far up in the woods and explaining how it had helped him to determine the highest point of sea level in Gwaii Haanas. Josenhans was there as well. "Hey," Josenhans suddenly piped up, "if you look carefully over there to the left, you can just make out Tom, poking his head out of the woods."

CHAPTER EIGHT

Ancient Mariners

THE PRETTY TEENAGER with a trace of Polynesian features danced a sultry hula to the music of a Hawaiian slack-key guitar. But this was not Oahu or Maui. It was outside the Maritime Museum on the waterfront in Vancouver, B.C. In the summer of 1995, two Hawaiian oceangoing canoes with native crews wowed Seattle, Vancouver, and every other port they visited up and down the West Coast, from California to Alaska. To save time, they had been shipped from Honolulu by freighter. But both of the twin-hulled boats had made epic Pacific voyages under sail, reenacting the original oceanic journeys that brought the Polynesians to Hawaii some 1500 years ago.

I had just published a book on the important role played by young Hawaiian men in the nineteenth century fur trade on the Northwest Coast. These "Kanakas" (simply the Hawaiian word for "person") were the largest ethnic group at the network of Hudson's Bay Company forts and outposts. Many of them ended up marrying native Indian women and settling in what would become Oregon, Washington, and British Columbia. Their part-Hawaiian, part-Indian descendants revered that heritage, and a few families still held luaus on local beaches.

Naturally I was excited by the canoe visits. So were the region's media, which gave the events extensive coverage, the TV cameras zeroing in on swaying hips in skimpy grass skirts. One theme kept cropping up in reporters' interviews with the Hawaiian sailors: astonishment at the very idea that small, open boats could ever have traveled such vast distances. I had noticed that most anthropologists, too, had a highly exaggerated fear of the sea, which might account for their resistance to the coastal migration idea. Like the reporters, the anthropologists tended to be landlubbers, a word derived not from "loving" the land but from *lober*, the Middle English for an awkward fellow or oaf, which sailors then applied in disdain to the landsmen who were so clumsy onboard a ship. No one captured this division of worldview better than one of the suntanned canoe captains. "So many people think of the ocean as a barrier," he told me, bemused. "We Polynesians have always thought of it as a highway."

Pacific islanders had developed very sophisticated forms of seamanship. Their seaworthy double-hulled canoes, or single-hulled ones with smaller outriggers, could skim effortlessly atop the waves. They even sailed well to windward, better in fact than European or American boats until the late nineteenth century. Borne on those graceful craft, the ancient mariners conquered half the world's southern oceans, from Madagascar to Easter Island, but this was only in the last two millennia.

The earliest arrival of people in the islands of Polynesia is well documented. They not only left archaeological artifacts, they also caused the abrupt extinction of many native animals and birds, and simultaneously introduced new species to the islands they reached. Until then, Pacific peoples apparently lacked the ability to venture across such vast stretches of open sea. So, just what were their capabilities prior to that time?

Scientists weighing the likelihood of late Ice Age coastal migration needed to know how advanced boating and navigational technologies were 10,000 and more years ago. Could Pleistocene migrants have made the lengthy crossings required to jump from one offshore island to the next along a stepwise journey through thousands of miles of coastal waters? Although ancient wood or skin boats, or more likely rafts made from logs or bamboo, were unlikely to be preserved in archaeological sites, anthropologists could nevertheless infer evolving boating technologies from the first arrival of people and their artifacts at various islands.

The best evidence for advanced maritime abilities in the late Pleistocene came from the southwestern Pacific. Greater Australia, which at lower sea level included the huge island of New Guinea, was colonized around 50,000 years ago. People migrating from Southeast Asia would have had to make a voyage, probably jumping off from Timor, of forty to fifty-five miles. From the combined Australia–New Guinea landmass, people next moved across deep (but not quite so wide) channels to the very large islands of New Britain and New Ireland in the Bismarck Archipelago of northern Melanesia. The latter was reached by 32,000 or 33,000 years ago, and the required crossing to get there was a bit less than thirty miles.

Although the distances are considerable, those colonizations are not proof of great seafaring skill. They could have been accidental. Anthropologists have sometimes quipped that a single pregnant woman drifting on a log could inadvertently populate an entire continent. That is probably wrong, and not only because it's almost impossible to cling to a wet rolling log. Try it some time, and imagine doing it in shark-infested waters! Population biologists and mathematicians have calculated the number of individuals required to implant a successful new human population. At least several fertile adults of each sex would almost certainly be needed, which implies a fairly large boat or raft, or else repeated transfers of people.

For the initial colonization of Australia, though, rather minimal boating ability might have sufficed. The continental landmass facing across the water to Timor was so large that the first peopling could have resulted from a number of accidental drifts on very crude rafts. Or it might have been intentional without requiring great navigational daring. Fifty or so miles puts Greater Australia over the horizon from Timor, but smoke from natural fires—caused, say, by lightning—could have given people on Timor secure knowledge that there was land beyond the sea's visible rim. This would greatly reduce the risk of setting out in primitive watercraft. They could have awaited ideal winds and currents.

Until the 1980s, the jury was still out on just how early people developed truly capable watercraft and navigational ability. But archaeology has since clarified the picture.

The next migrations in the southwestern Pacific, onward from New Guinea to New Britain and New Ireland, were almost certainly intentional. Obsidian, a dark, glassy rock formed from silicon at Hadean temperatures inside volcanoes, is ideal for making viciously sharp tools, and the specific volcanic source can be determined by laboratory analysis. Obsidian from New Britain was carried to New Ireland, where it turned up in sites some 15,000 to 20,000 years old. There would have been a water gap of around twenty miles to cross. Small marsupial animals were also taken from New Guinea to much smaller islands in the region. Presumably this was deliberate, to stock those islands with succulent beasties for future hunting. Still, New Britain and New Ireland are very large and mountainous islands. Reaching such visible target destinations did not require striking out into the unknown.

Moving onward from New Ireland and colonizing the Solomon Islands required a much longer crossing. Archaeologists specializing in that region had long assumed that the Solomons were colonized only a few thousand years ago. Then, in 1987 Stephen Wickler, a graduate student at the University of Hawaii, spent eight months digging in limestone rock shelters and caves on the eastern side of Buka Island in the Solomons. He was searching mainly for shards of the telltale Lapita pottery, a type of geometrically decorated fired clay that spread rapidly across the southwestern Pacific beginning around 3500 years ago, and has been used by archaeologists to trace the people who made it. Until Wickler's effort, no known human occupation on Buka or the much larger adjacent island of Bougainville predated about 2500 years. What he found was a total surprise.

While excavating a trench in shallow Kilu Cave, overlooking the sea, Wickler found well-stratified deposits more than six feet deep. The upper levels did, indeed, contain pottery, as expected. The lower levels had no pottery, but they were full of mammal, lizard, and fish bones, as well as shells, fire hearths, and flaked stone tools. Radiocarbon dating of the shells, which were in close association with the stone tools, indicated that the people had reached the site as early as 29,000 years ago. This was a shocker.

And then, another surprise: people back then had not lived only on fish, shellfish, and seaweed.

Thomas Loy, a researcher based in Australia, specialized in detecting and identifying microscopic particles of organic matter clinging to ancient tools. (The world of paleoanthropology is small. During that same period, Jim Dixon was working with Loy to study blood residues on stone tools found in Alaska, attempting to detect mammoth blood by its hemoglobin.) Most of the Buka stone tools had grainy residues on them from taro, a starchy root that has long been one of the South Pacific's most important staple crops. There were also starch grains from a similar plant that could not be identified. This trace of taro was the earliest direct evidence for the prehistoric use of root vegetables anywhere in the world.

Wickler's discovery was a real eye-opener for anthropologists, such as Geoffrey Irwin of the University of Auckland, in New Zealand, who were trying to understand ancient seafaring capabilities and strategies. Irwin himself was a blue-water sailor who had voyaged offshore not only in his own sloop but in traditional native canoes as well. He applied his experience to prehistoric Pacific migration and colonization.

Irwin realized that the height of local land forms had to be plugged into the equation when assessing how far from shore a boat could journey safely without navigational aids, and nothing like a compass or sextant would exist for thousands of years. From a rowing boat or raft, a low atoll, even one with tall palm trees, cannot be seen from more than about ten miles away. Mountainous islands, though, are visible from much farther. Irwin took the nautical charts of the areas he was studying and drew circles of visibility on them that varied with the height of those islands.

As the crow flies, Buka lies 110 miles away from New Ireland, which was settled by 32,000 years ago. That put it far over the horizon for anyone setting out from New Ireland, as they must have done, 29,000 years ago. Along a circuitous jog from New Ireland to Buka, though, lies the island of Feni, and farther along that curved route is a group of low atolls, the Green Islands. The geology of the Green Islands was uncertain. Coral atolls grow slowly with the buildup of untold trillions of minute calcareous skeletons. If the Green Islands were already visible above the waves at the time of Buka's earliest colonization, then the open sea crossings required to hop from New Ireland, to Feni, to the Green Islands and finally to Buka, would have been thirty to forty miles.

Such distances would be no greater than other crossings people had made, moving from one very large or high island to another; however, even if the small and low-lying Green Islands did exist as possible way stations, they would not have been visible from Feni. They could only have been discovered by an intentional search beyond the horizon, or by an accidental drift. And if the Green Islands were *not* available as an intermediate stepping-stone, a crossing of almost ninety miles would have been required to get from Feni to Buka.

Irwin analyzed many such crossings and early colonizations in the region. Tracing arcs on his charts, he was able to determine the limits of visibility relevant to getting from one island to the next, and to define a situation he called "two-way intervisibility," where the distances and height of land were such that both the jumping off island and the destination island remained visible all the way. With longer distances and lower islands, though, there might be only "one-way intervisibility." This meant that the mariner had to set out without the target land in sight. If the jumping off island were mountainous, though (like New Ireland), a raft or canoe might be able to proceed quite far offshore before losing sight of it. Meanwhile, unknown land ahead might appear over the horizon before the high land behind faded from view. This could happen while a boat was on an intentional search for neighboring islands, but also accidentally, while simply out fishing.

Finally, there was the situation of no intervisibility, where a mariner would have to lose all sight of land and where, before compasses, it would have been a challenge to make it back to the jumping off point if no other land came into view. At first glance, it might seem that anyone voyaging over the earth's watery curve in these circumstances would be making a wild, life-threatening roll of the dice. If the sailor did not happen upon a new and unknown island within a reasonable distance, he could try turning for home, but all he would see was a blank horizon. How would he know the right direction to steer, and how could he maintain a steady course even if he knew? But in fact—and here's where the landlubber's fear of the sea is based on lack of knowledge—he would not be totally blind or beyond hope. A lifetime of experience with the prevailing winds and wave patterns in the region would help. Polynesian islanders have navigated that way for millennia. Observing the flight of shore-based birds also provides clues to the direction of the nearest land, and might even have helped ancient navigators to find new, unsuspected islands. In addition, the tropical sun rises so high in the sky that it is useless in judging direction for much of the day, but at dawn and dusk it hovers low over the horizon as a guiding beacon.

For Irwin, the most dramatic example of a no-intervisibility colonization in the southwestern Pacific was Manus Island in the northern part of the Bismarck Archipelago. Manus is about 125 miles from the nearest jumping off point, and it has a relatively low elevation that would require a blind crossing during which voyagers would have been out of sight of land for a minimum distance of forty to fifty-five miles, depending on the direction of approach. The dating of archaeological finds on Manus was not as firm as for Buka, but they were certainly no more recent than the very late Pleistocene. The evidence for this was that by 12,000 years ago, obsidian from Lou Island, about twenty miles to the south, was being brought to Manus. And there were transfers of *Canarium* nuts and an animal species, the bandicoot, all the way from New Guinea as well.

Irwin concluded from his studies that there was at least one *definite* late Pleistocene example of navigating far over the horizon (Manus), and one much earlier *likely* example (Buka), depending on whether or not the Green Islands were available as low, difficult-to-spot stepping-stones. Irwin further showed that the temporal order of island colonizations was far from random. In the southwestern Pacific, the islands that had two-way intervisibility were consistently colonized earlier than those with one-way intervisibility, and those with no intervisibility were colonized last. In short, there was a logical and systematic pattern to the timing of these ancient population movements. It was skill, not blind luck that had led to these island colonizations.

And the southwestern Pacific, with its tropical climate, was by no means unique. Offshore voyaging did not require the coddling comforts of warm water and balmy breezes. On the much colder northeastern coast of Asia, advanced boating skills must also have existed by the late Pleistocene. By the early 1990s, archaeology showed that obsidian from Kozushima Island, south of Tokyo, had reached huge Honshu Island by 32,000 years ago, which required a deepwater crossing of around twenty miles. The island of Okinawa was also probably colonized about as early, and certainly by around 18,000 years ago, a movement of people that required multiple island hops.

If people in both the southwestern and northwestern Pacific had capable watercraft well before the end of the last glaciation, there was little reason to doubt their ability to move in a long arc around the North Pacific rim. Such a coastal migration would mainly involve relatively short hops, allowing for frequent stops to seek safety from the weather, to rest, and to resupply. Much of the coastline is mountainous, making it visible from a considerable distance offshore, at least when the weather was clear. Even if migrating people sometimes lost sight of the shore, their situation would be far from dire. For major stretches, such as the coast of Kamchatka and most of the Gulf of Alaska, the shoreline is a long, unbroken expanse of land. The "target," as with the coast of Australia, is almost too large to miss. Even the roughest sense of direction, determined from prevailing winds, swells or winging birds, or from the sun in the sky, would allow mariners to head for shore and safety.

The one question mark that hung over this otherwise favorable assessment of ancient coastal hopping—and the reason for some legitimate skepticism about it on the part of anthropologists—was ice. Well into the 1990s, the accepted wisdom was that access to much, and perhaps all, of the late Pleistocene North Pacific coast would have been blocked by sheer cliffs of glacial ice fronting the sea. As with the supposed ice-free corridor, this conclusion was based on minimal (and in retrospect very shaky) evidence from geology.

There was no doubt that the coast had been impacted by glacial ice, but I was appalled to learn how little of our supposed knowledge was based on solid research. The timing of the advances and retreats was hard to pin down, and the necessary scientific sleuthing had simply never been done. In many places, glacial flows had terminated offshore, below today's sea level, which made the marginal features of the glaciers, such as the moraines, inaccessible. Most authoritative glaciology maps of the Alaska coast showed all of the Aleutian Islands, the Alaska Peninsula, Kodiak Island, and the Gulf of Alaska totally covered with ice well out to the edge of the continental shelf. If valid, such a long stretch of totally icebound coastline would indeed have hindered coastal migration. It was hard to imagine Pleistocene people bobbing along on the swells for weeks at a time past a hostile and unbroken barrier of ice. Understandably, archaeologists accepted the published findings of glacial geologists. Most concluded that movement by people around the North Pacific rim was impossible until the glaciers retreated fully. The timing of this retreat was unknown, but generally thought to be quite late.

New research changed this picture. Alaska-based geologist Daniel Mann and several colleagues had been studying and dating stratigraphic layers on land since the mid-1980s. The goal was to unravel the exact sequence and chronology of the last glacial maximum on the Alaska coast. They analyzed pollen from sediment cores and glacial till on Kodiak Island and the Kenai Peninsula, and were surprised by what they found.

In 1994, Mann and Dorothy Peteet published a study showing that even at the very peak of glaciation, roughly 17,000 to 20,000 years ago, there had been large ice-free areas on Kodiak Island and at least one such area on the Alaska Peninsula. On its own, the existence of a couple of ice-free refuges might not prove much. Mann's results also showed that most of the rest of the potential migration route *was* covered with glacial ice at the peak of glaciation. "Our reconstruction," Mann and Peteet wrote, ". . . suggests that the [Alaska Peninsula Glacier Complex] covered most of the continental shelf in the Gulf of Alaska around Kodiak Island and the Alaska Peninsula." This would have presented a substantial barrier to coastal migration. But Mann was also able to determine the timing of the retreat of the glacial ice quite precisely. The calving ramparts of blue were melting rapidly by 16,500 years ago, and many areas along the relevant coastline became ice-free starting around 16,000 years ago. At any time after about 16,000 years ago, therefore, and almost certainly by 15,000 years ago, maritime people should have been able to coastal hop around the North Pacific rim without making long offshore jumps. Mann's research "opened up" what had been considered the longest and most difficult part of the coastal route from northern Asia into the New World.

The prospects for coastal migration to the east and south of the Gulf of Alaska also began looking much brighter in the early 1990s. As Heiner Josenhans pointed out during the *Vector* cruise, seafloor sediment cores had brought up wood, pollen, conifer needles, and other evidence of late Ice Age refuges on the northern B.C. coast. On a 1991 cruise to Dogfish Bank in Hecate Strait, for example, Vaughn Barrie obtained three cores containing "terrestrial sediments representing a tundra environment." The dates were around 13,790 years.

Rolf Mathewes of Simon Fraser University studied those cores. They confirmed what he had already found from earlier analyses of pollen and fossil plant debris on land in the northern Queen Charlottes, that there had been several huge and conveniently spaced offshore refuges to the east and south of the Charlottes that would have been free of ice by around 14,000 years ago.

In colorful lectures documented by slides, Mathewes proclaimed that these places represented an entire "lost world" that was subsequently drowned by the rising sea. People moving by watercraft, he argued, could have found edible herbs, starchy plants, and berries on land. Along the sea's edge there would have been abundant shorebirds, shellfish, edible seaweeds, and sea mammals to hunt. The tundra itself would have been populated by caribou, bison, and possibly even mammoths and musk oxen. "Here, we have an Eden-like environment that had food resources and fuel and medicines and whatever else people would have needed," he added. "When you compare that to the ice-free corridor that has always been favored by the archaeologists, this area looks a lot more passable and a lot more benign."

Another significant refuge was the Brooks Peninsula, a rugged finger of land that juts out from the coast of northwestern Vancouver Island. Biologist Richard Hebda of the Royal B.C. Museum collected pollen and other organic samples there and had them dated. His conclusion: Glaciers apparently covered the peninsula during the Ice Age peak, but then it became ice-free "likely by 13,000 to 14,000 years ago, and maybe earlier."

Proponents of coastal migration were heartened. The archaeological mainstream, though, was less impressed. Those who were most skeptical of coastal migration and most committed to Clovis First ignored the emerging findings entirely. And even when they did not dismiss it outright, there was at best an academic time lag of about five years before the new evidence was absorbed.

I had to shake my head in dismay at the stubborn, hidebound resistance to the emerging geological and glaciological picture, and to coastal migration itself. "I've got a hard time with boats and refugia," said one very prominent archaeologist, David Meltzer of Southern Methodist University, in 1998. "You have huge chunks of ice, hundreds of meters high," and they're "coming straight down and calving into the ocean. And there are a whole bunch of them." Meltzer adopted a worst-case image of these shoreline glaciers creating their own nasty localized climate, with powerful winds and floating bits of broken ice. To him the very idea of coastal migration was almost laughable. What's going to grow on these glaciers for people to eat? he asked rhetorically. "Are you going to have these cute little 7-Elevens?" he asked in ridicule. Places where you could buy food or drink "your 'Big Gulp' and make your way down south?"

Meanwhile, on the northern B.C. coast, the scientists continued their basic research. In 1995, Josenhans, Barrie, and Fedje were able to get ship time on the large *John P. Tully* for a survey and coring cruise in Charlottes waters. It was capable of carrying the bulky vibracore system needed to penetrate the hard sediments that had so frustrated them on the *Vector* cruise.

They obtained several cores in northeastern Hecate Strait that confirmed Barrie's 1991 findings: toward the mainland side of Hecate Strait the ice had apparently retreated by 14,500 years ago, and the land quickly rebounded, leading to a rapid drop in sea level. There were indeed places to the east of the Charlottes that were dry and ice-free tundra by around 14,000.

And they located additional areas to the east of Gwaii Haanas, in the southern portion of the Charlottes, that had been high and dry at the end of the Ice Age. The most important one was Middle Bank, which is southeast of the southernmost tip of the Charlottes. It was an ice-free island that apparently remained dry land until the rising sea inundated it between 11,000 and 12,000 years ago. That added another conveniently located stepping-stone for any people who might have been moving along the coast prior to that time. In all, therefore, between the Queen Charlottes and Vancouver Island there were three large late Ice Age islands (Middle Bank, Goose Bank, and Cook Bank) each of them at least one hundred square miles in size.

The network of known offshore refuges was becoming denser, the distances between them much shorter. Logically, coastal migration seemed increasingly credible, but when I pondered it, even I had nagging doubts. I had been hundreds of miles offshore on a small sailboat, but only in tropical waters. I had even built and repaired boats myself, but it was with modern tools and relying on thousands of years of accumulated knowledge. Like the skeptical landlubber archaeologists, I still found it hard to imagine building boats or rafts with stone tools. And I was haunted by images of navigating well off the coast on crude watercraft in storms and freezing conditions.

My doubts dissolved, though, during the flight I took from Seattle to Ketchikan in the summer of 1999 en route to Prince of Wales Island for my visit to On Your Knees Cave. I was lucky to get a window seat on the left-hand side of the Alaska Airlines jet, facing west. It was a perfectly clear day during one of the glorious midsummer North Pacific highs that often linger for a week or two. I kept my forehead glued to the window, while a panoramic view of the formerly glaciated Northwest Coast unrolled below like a slow conveyor belt.

As our 737 jet reached cruising altitude, the snowy peaks of the Olympic Peninsula in Washington State gave way to the snowfields and small remaining glaciers on the higher mountains of Vancouver Island. The lower areas were an emerald carpet dotted with lakes. Like a video in fast rewind, my imagination scanned backward through hundreds of human generations. In place of today's lush green world was one in which vast flowing lobes of ice bulldozed their way down off mountain ridges and splayed out into the lowlands. Nothing showed above the sheet of white but barren and lifeless nunataks, high sterile peaks whipped by vicious winds. But off to the west, the Brooks Peninsula, protruding from Vancouver Island out into the Pacific, would have at least offered the refuge of a tundralike seashore rich in marine resources by around 14,000 years ago.

The plane left Vancouver Island behind. About fifteen miles north was the area of shallow sea that, I knew, covered Cook Bank. In the distance I could already make out the dim outline of the Queen Charlottes. Nicely spaced in between, I could imagine the locations of Goose Bank and Middle Bank, both of them large, ice-free islands in the late Pleistocene. In my mind, I peeled away 300 or more feet of water. A boat or raft traveling south from the Charlottes to Vancouver Island would not have had to make any jumps longer than about twenty miles.

During the last half hour of the flight, as we approached Ketchikan, I could pinpoint the undersea areas to the east of the Queen Charlottes, Laskeek Bank and Dogfish Bank. Here, again, no long jumps would be needed to go from one refuge to the next.

To the north, the first islands of Alaska appeared across placid Dixon Entrance. Barely a wisp of wind rippled the water below. Unlike my first crossing of Hecate Strait in a southeasterly gale, on this day it was like the proverbial millpond. The distance from the Alaskan islands to the huge, high, and visible Queen Charlottes was only about thirty miles, a mere stone's throw. It was hard to imagine the crossing taking more than a couple of long summer days (and very short high-latitude nights) even paddling the crudest raft, dugout, or skin boat. By the time we landed in Ketchikan it had really sunk in. For mariners with the sense to wait for summer and a spell of good weather, traversing the coast should have been a piece of cake.

Still, only archaeology could prove they had done so.

CHAPTER NINE

Archaeology Gets Its Feet Wet

WADING OUT INTO the shallows, Kim Conway zipped up the back of Pete Waddell's dry suit, and Waddell returned the favor. They checked each other's air hoses and regulators, inserted their mouthpieces and ducked their heads under to make sure their air supplies were working. After a quick thumbs-up, they sloshed out farther and slipped from sight. Parallel lines of bubbles showed their location as they swam straight out from the beach and made a few circular sweeps one or two hundred yards from shore. They angled over to the rocks and kelp on one side of the small cove, then turned and headed back the other way until they ran into the shallows again. Deckhand Rob White stood poised, ready to push off with the Boston Whaler, in case they needed help.

This first undersea reconnaissance of the *Vector* cruise was aimed at checking out the sea bottom close to one of Daryl Fedje's most productive onshore archaeological sites, Echo Harbor on Moresby Island. Fedje had asked the divers to size up the potential for a future underwater dig. He needed to know how the bottom compared to what he could see on the beach, especially in shallow areas where divers could work for extended periods without risking the bends when they came back up.

After only fifteen minutes in the water at Echo Harbor, Waddell and Conway frog-marched up the beach in their fins. Then they repeated the operation across Darwin Sound at the Richardson Island site. Following a brief dive and sweep back and forth, just off the beach where Fedje and McSporran had found thousands of artifacts earlier that year, the divers briefed Fedje on what they had seen.

It looked very doable, said Conway, especially off Richardson Island. The bottom there was gravel and full of clearly visible cobbles at least down to fifty or sixty feet. It was what geologists call a lag surface, where the fine sediments have been winnowed out (the opposite of being silted up, which would hide any artifacts). The strong currents, flushing through Darwin Sound parallel to the shore, were keeping the seafloor free of sediment.

"Look at how Daryl can just walk along the beach and find the remnants of ancient stone toolmaking," Conway told me. "Someone with training could do the same underwater." Even a systematic excavation could be carried out. *You'd need many hours of dive time*, he thought. And the diver would have to be shown by an archaeologist exactly what to look for in the way of artifacts. But then, he added, "If you had a small group that could work right on the beach, with a small compressor to refill their tanks with air. And if you set up a camp, so you could stay maybe two weeks. I think you'd have a pretty good chance of success. Look at Montague Harbor," he went on, referring to a sheltered site on Galiano Island in B.C.'s southern Gulf Islands, where an underwater archaeological team had done several seasons of undersea digging. "The same kind of technique could be applied here," at least in the shallower areas. "You have unlimited dive time at thirty to forty feet."

When I got home from the cruise I phoned Norman Easton, a young archaeology instructor at Yukon College, and asked him about that landmark dig. Easton had been impressed by Knut Fladmark's arguments for coastal migration, but he had not shared Fladmark's pessimism about the prospects for finding undersea evidence. Fladmark was not a scuba diver; Easton was. His idea was that if archaeologists really wanted to investigate coastal migration, he and his colleagues would have to lay down their trowels and take up regulators, tanks, and diving suits. By the time Easton got going in the late 1980s, there had been lots of underwater archaeology, but not the kind that was needed to find traces of ancient coastal people. "The difference," he told me, "is that most underwater archaeology to date has focused on structures of one sort or another; principally shipwrecks, and to a lesser extent monumental architecture, where the excavation is oriented to large, built objects." Nearly all of it was from historic times. By contrast, "prehistoric evidence is going to be at the relatively micro level of smaller artifacts and other types of environmental indicators."

Diving on shipwrecks or large structures was a situation where there was usually only a single date involved, in many cases known from historic sources. A wreck was like a snapshot, a single moment frozen in time on the day the ship took its plunge to the bottom. Keeping track of the stratification of different layers in the underwater site was not critical, because everything found was presumed to be of the same age. Unlike most land digs, shipwreck archaeology did not mainly require the sequential peeling away of layers of sediment or coaxing elusive snippets of information from tiny objects. In that sense, Easton's project was unique.

He had chosen Montague Harbor because it was extremely sheltered, an ideal hurricane harbor completely surrounded by land. On shore there were rich deposits of prehistoric material in the beaches, leading right down into the lowest low tide zone. These had been thoroughly studied in the 1960s during a dig led by Donald Mitchell of the University of Victoria, which found artifacts going back about 3000 years. Given local sea level change, anything older had to be underwater. The task was to extend the archaeological record out from those rich beach sites into the shallow waters of the harbor.

Easton had to squeak by on a tight budget, renting a cheap house on Galiano Island for his crew accommodations and recruiting a team of divers willing to work for free. Fortunately, there was no shortage of volunteers; eager recruits came from as far away as California and Alaska. On average, there were up to fifteen people on site at a time, and eighty to a hundred in all in the course of a summer. Their first reconnaissance dives were in 1989, and they got down to serious, systematic work over the next three years. Because the water was only about twenty feet deep, most divers put in three hours a day, divided into two shifts.

It was all trial and error. At first they tried to dig their way down into the bottom sediments without any structure around the excavation, but the sides of the hole tended to collapse. Material from above filtered down and in, so in 1991 they set up a square underwater caisson, a boxlike unit about eight feet along each side, with the top and bottom open. Excavating within it prevented the unwanted intrusions. The divers used trowels and an airlift system to bring the excavated material to the surface. As they cleared away the sediments they examined them for artifacts. The remaining sediments were brought up from the bottom by the airlift, which was like a big vacuum cleaner. Fed first into a holding basket, the diggings were later carried ashore for screening, as at any other archaeological site.

Easton's team measured precisely and kept track of the location and depth of everything they found, just as they would on land. "So we know which little square the sediments came from, after screening. It's all plotted on a chart, so it can be reconstructed." As they worked their way deeper into the sea bottom, they put a new caisson unit on top of the first one and dropped the entire structure down. When they stopped for winter, they covered the caisson with a tarpaulin, then went back to it the following year. In all, they dug at three different places in the harbor.

By the time they ran out of money, the team had reached the bottom of the deposits in one spot. They obtained radiocarbon dates going back 6700 or 6800 years and found lots of artifacts, including chipped stone points such as a little arrowhead. There were also lots of bone artifacts, including their most elegant find, a very finely made harpoon head of antler that appeared to be at least 6700 years old. The dig had been a great success.

"The next place we'd like to go is the west coast of Vancouver Island," said Easton. But he knew how hard it would be to attempt similar work in more exposed and rougher places. "Locations more directly related to the coastal migration route are going to be in much deeper waters and wilder areas," he admitted. Working far from shore, from a boat or ship, would be very expensive. And digging in really deep water, where the older stuff was, would drastically limit dive time and require slow ascents to the surface. Easton never did repeat the effort. Soon he went back to graduate school to complete his Ph.D.

Daryl Fedje also realized that using divers to dig in water hundreds of feet deep was not practical. Submersibles could stay down for long periods, but if they carried crews, they were too expensive to be a realistic option. Robot submersibles were much cheaper, though; their capabilities for marine geology and archaeology were worth considering.

The second working day of the 1994 cruise, a Parks Canada boat rendezvoused with the *Vector.* On board was technician Henk Don, head of diving operations at Canada's Center for Inland Waters in Burlington, Ontario. Don had a background in civil engineering and did most of his work in lakes and rivers. He was the designer, keeper, and doting godfather of a compact 170-pound robot submersible called "MURV," which stood for Mobile Underwater Reconnaissance Vehicle. From the start, Don talked about MURV as if it were an incarnate being with personality. MURV was a feisty but occasionally temperamental little guy who had to be coddled just a bit. What he lacked in size and power, he made up for in patience and obedience to commands. Josenhans used more generic marine terminology, calling the diminutive beast a "ROV" (for Remotely Operated Vehicle). He even turned the initials into a verb, as in "ROVing" across the sea bottom.

The ship's crane gently hoisted the sturdy blue containers holding MURV and his auxiliary equipment onto the deck. Everyone crowded around to gawk as Don undid the latches on the main container to reveal an ungainly looking black-and-yellow contraption about the size of a large portable generator. Emblazoned on the little robot, along with "MURV," were "Canada" and the Maple Leaf flag. There were flotation chambers to control buoyancy and electric motors to provide thrust. The device was also equipped with sonar, a still camera, a video camera with lights, separate strobe lights for the still camera, hydrophones to home in on beacons, and a little mechanical arm. A lot of gear in a little package.

MURV had not been purchased off the shelf. The core of the unit, especially the flotation and electric motors, was built in Massachusetts, but then Don tacked on components to suit his needs. "We added the little robot arm," he said. The sonar was extra, too. MURV was designed to work to a depth of 1000 feet and was linked to the ship by an umbilical cord 1150 feet long. Electrical power (and commands) flowed down the cord from the mother ship to run the motors and electronics. Signals shot back up the cord to a veritable boggle of instruments and controls. With the monitors, sonar, and other bells and whistles, the little sub had cost around $100,000.

But it was money well spent. MURV was often used in deepwater situations involving search and recovery. "A lot of equipment falls off ships," Don explained. "Expensive stuff. For example, a current meter on a mooring would cost us $18,000, so we put an audio pinger on the meter to give us a homing beacon. And we put a hydrophone on MURV to home in on it." MURV was also used in programs studying fish habitat. "You put MURV down on a spawning shoal, he sits there quietly and doesn't scare the fish, but he records the lakers coming in and spawning in the night." Even when using human divers on a project, MURV was often deployed for safety and convenience. "Our divers love being in the water with this guy. MURV can watch with his cameras, thereby providing security, and he also provides light, so the diver doesn't have to hold lights himself. It's a very safe way to work." Don had used other robot submersibles in the past, but they were big and heavy and cumbersome. This one was nicely miniaturized. "We've been working with this little bug here since 1985."

Josenhans had hopes that MURV could help him view and understand the geology of the area. Tomorrow, he said, "we'll be at Matheson Inlet, where there is an ancient river channel. What we're seeing there is bedrock, with old lake sediment draped on top of that." A steep cliff had formed with horizontal stratification, and Josenhans wanted to get a close look at it. "We'll just let MURV climb up the cliff," sending back pictures.

To show me another place where MURV might be useful, Josenhans pulled out a sidescan sonar printout and spread it on his chart table. "That light spot is the soft sediment that would have been the old river channel. And here—a bedrock outcrop—this would have been the high ridge that constrained the river valley. This is where we'll do the core, right there. See the soft sediments there, snaking off into Echo Harbor? We'll ROV up the flanks of this thing later in the day and zigzag in toward shore. Beautiful! That's where we want to go, right there." He pressed his finger down on the paper.

MURV was first deployed the next day, when there was a lull in the coring action off Echo Harbor. "Let's put down MURV," Josenhans announced. Pete Waddell took charge of coordinating the procedure on deck, where deploying MURV required teamwork. To have the delicate instruments out of the weather, Don had set up his video and sonar monitors and controls inside the lab, where he could sit in front of them in comfort and steer MURV with a joystick. Waddell could communicate with Don by walkie-talkie and talk to the officers on the bridge as well. He could also give hand signals to Roger LaVigne, the bosun, who was perched high above on the second deck, where he used the ship's crane to lift and lower MURV over the side. The 1100 feet of umbilical, kept on a big spool, had to be paid out in a controlled fashion. If too much slack were allowed in the line, it might snag on the bottom. Worst of all, it could get sucked under the ship and into the propeller.

Don was concerned about the current, and with good reason.

MURV's top speed under power was only two knots, while bottom currents often whipped through the islands at least that fast. They simply overwhelmed MURV on the first attempt. He was lowered into the water, releasing a burbling fountain of air bubbles that moved away from the ship. But the currents were much too strong. MURV was swept off sideways and into the depths, out of Don's control. I was concerned about the little guy's fate, until I remembered that he was firmly tethered by the umbilical, which the deck crew used to retrieve him.

Later in the day, there was another lull in the other research tasks, so they tried again. "Dive, dive, dive," Waddell chanted like a mantra as MURV slowly sank and disappeared from view. Most of the team from the aft deck slipped into the lab to peer over Don's shoulder at the monitor and controls. He stared intently at the colorful, multisectioned video monitor. One part of the screen displayed a compass that showed MURV's direction of movement. There was also a readout for depth. "I'm at the sixty-five-foot mark," Don told Waddell over the walkie-talkie. "He's drifting back more under the ship than he should." But Don decided to go deeper. "Proceed to twenty-five fathoms, Peter," or 150-foot depth. Then he took MURV down farther still. When 300 feet of umbilical were out, we knew we should soon be seeing the bottom. MURV's lights were on, but all we could make out was a lot of bright, white flakes drifting past, just like snow. "It's a real organic soup down there," said Josenhans.

White, the deckhand, loved fishing. "Can't you turn the camera so we can check whether there's any halibut or dogfish down there?" he joked with Don. "We could jig for it right over the side."

The video camera could see no farther than the water clarity and lights from MURV permitted. But the submersible also had a sonar system that could image the sea bottom and anything large on it out to a hundred yards or more. "We should be able to see the walls of the channel," Josenhans remarked. But so far the sonar showed nothing. Then, an image began to resolve itself on the video screen. It was a round white object, a shell. MURV was hovering just over the bottom at last. The readout said 276 feet. Other than the shell, there was not much to see at first, only bubbles drifting past and some fishes moving around. It was hard to judge the size of the fishes as they darted in and out of the glare of the lights. Don worked the joystick and set MURV down on the seafloor. Briefly a cloud of fine stuff blurred the view, then the current whisked it away. The current was so swift that it looked as though we were still moving, plowing ahead. Organic particles and bubbles swept past the camera.

Next Don used the submersible's small thrusters to bring MURV up off the bottom and swung him around, now pointing east. In the light of the video camera, we could see the tracks on the bottom left by MURV's skids. But Don was having problems again. MURV was in the relentless grip of the current and the ship's drift, which tugged on the umbilical. "The second I try to maneuver, the slack gets taken up. I'm at full thrust right now. Peter, please give me another twenty-five feet," he pleaded into the walkie-talkie. The extra slack allowed MURV to steer again, and soon we were flying along over the bottom.

Into view came a sea fan, a filter-feeding animal that looked like a quill or feather, sticking straight out of the bottom. Then we came upon a flatfish (a sole or small halibut) that scurried away as soon as the lights shone on it. Next we could make out the sinuous tentacles of what may have been an octopus as we zipped past. There were starfish here and there. Still, the view was not very clear. Suspended in the water column was so much plankton, it was like driving through a snowstorm with the headlights on. At one point, the lights and camera zeroed in on a distinct dimple in the sea bottom. It was the hole left by a core that Josenhans had taken almost an hour earlier. Then MURV moved in close to a flatfish lying on the bottom. Don reached out with the robot arm and touched it. It jumped, scooted off a short distance and settled down again. MURV came up close again to tease the fish, but Josenhans had had enough of these games. "We're not going anywhere in a hurry," he said impatiently, and soon called off the dive.

The author carries a bucket of sea floor sediment that will be screened by an archaeologist for artifacts. On board the research ship *Vector* in the Queen Charlotte Islands off northernmost British Columbia, 1994.

Marine geologist Heiner Josenhans lets excess water drip from the dredging jaws before bringing onto the aft deck a large bite of the sea bottom that was once dry land, but now lies under hundreds of feet of water.

Josenhans studies his charts in *Vector*'s laboratory while planning where to take a core, which will give him a long, vertical sample of sea floor layers that reflect thousands of years of sediment deposition and allow him to establish the date of changing sea level.

Divers Pete Waddell and Kim Conway prepare to search sea bottom at Echo Harbor in the Queen Charlotte Islands, 1994.

Technician Ivan Frydecky deploys the sidescan sonar over the stern of the *Vector*, 1994.

Archaeologist Ian Sumpter watches as Haida trainee Tom Greene Jr. digs a piece of whalebone from the trench around a totem pole at Sgan Gwaii in the Queen Charlotte Islands, 1995.

Archaeologist Daryl Fedje supervises as his crew uses hand winches, or "comealongs," to pull a leaning totem pole into an upright position.

Haida archaeology trainees Sean Young (left) and Jordan Yeltatzie screen the totem pole diggings for artifacts and human remains, Queen Charlotte Islands, 1995.

Archaeologist Jim Dixon (left) and crew foreman Craig Lee inspect a meter-square excavation at On Your Knees Cave on Prince of Wales Island, Alaska, 1999.

Student archaeology volunteer Heather Mrzlack digs her assigned square of forest soil outside On Your Knees Cave.

Pete Smith, a Prince of Wales Island caver and contractor, drills holes and sets off tiny explosive charges to break up rock inside On Your Knees Cave.

Paleontologist Tim Heaton sifts through the previous day's diggings for tiny animal bones at his beach campsite on Prince of Wales Island.

A 9,200-year-old obsidian microblade unearthed by teenage volunteer Dylan Reitmeyer outside On Your Knees Cave.

Prince of Wales islander Bob Gray explores El Capitan, Alaska's largest known cave, which has thousands of feet of passages, a river, even tiny lakes. It also provided paleontologists with the first Ice Age bear fossils found on the North Pacific coast.

Flowing water in El Capitan carved out narrow, vertical "chimneys" that connect several levels of larger passages.

Image created by swath bathymetry (multibeam sonar) of the ancient drowned river delta site at Werner Bay in the Queen Charlotte Islands. A 10,200 year old artifact was dredged from the sea bottom at the indicated spot in 1998. Courtesy of Heiner Josenhans, Geological Survey of Canada, and Daryl Fedje, Parks Canada.

Decaying cedar totem poles line a small seaside forest clearing at Sgan Gwaii, the best preserved ancient Haida village in the Queen Charlotte Islands.

Commercial boat fishes for salmon near Kunghit Island in the southern Queen Charlotte Islands.

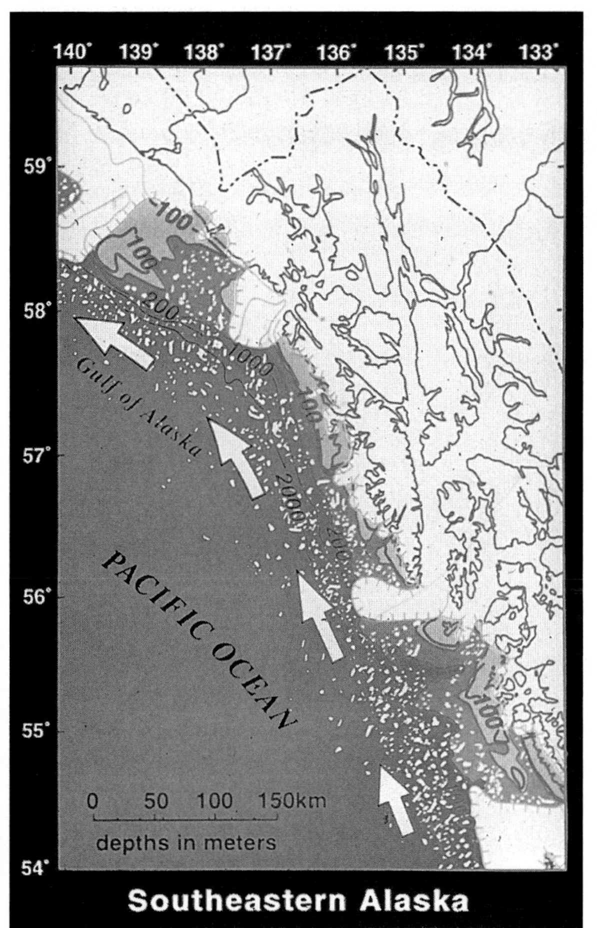

Glacial ice flows and ice-free refuges in southeastern Alaska at the peak of the last glaciation (around 18,000 to 20,000 years ago), superimposed on modern-day shorelines and water-depth contour lines. The ice-free areas are shown in a middle tone of gray. Courtesy of Tom Ager, U.S. Geological Survey.

Glacial ice flows and ice-free refuges in the Gulf of Alaska at the peak of the last glaciation, superimposed on modern-day shorelines and water-depth contour lines. The ice-free areas are shown in a middle tone of gray. Courtesy of Tom Ager, U.S. Geological Survey.

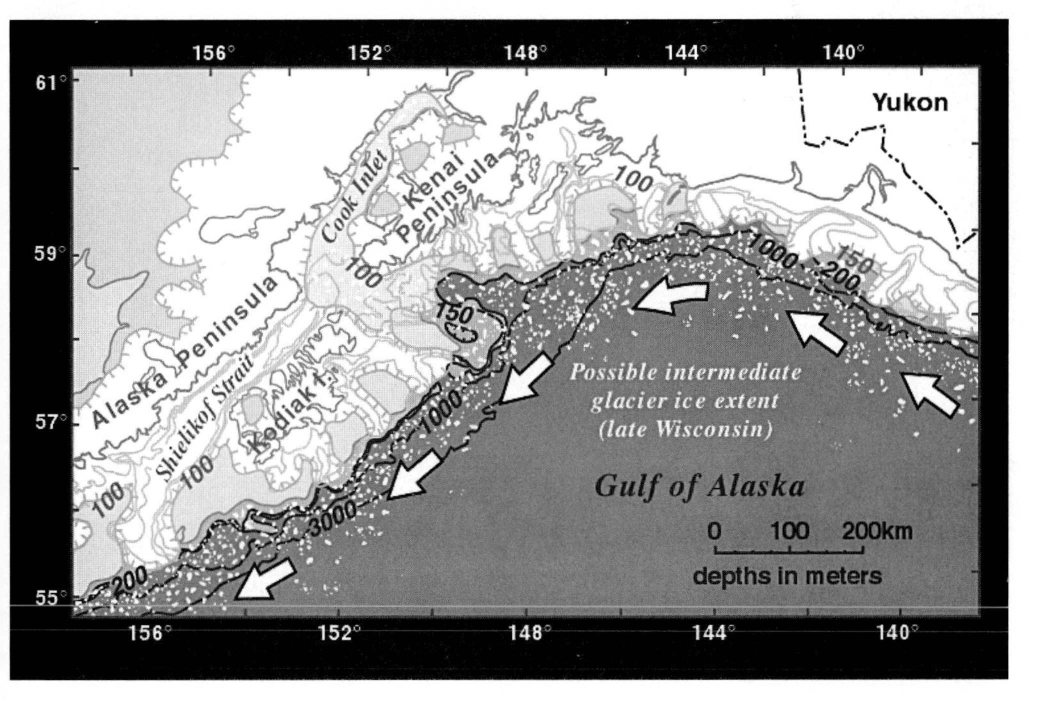

The basalt stone flake artifact dredged from the sea floor at Werner Bay, the oldest known artifact in the Queen Charlotte Islands. Courtesy of Joanne McSporran.

Phil Orr, who excavated human and pygmy mammoth elephant sites on Santa Rosa Island, California. He believed that people had inhabited California's Channel Islands 20,000 or more years ago. Courtesy of the Santa Barbara Museum of Natural History.

Orr found the leg bones of "Arlington Woman" protruding from a hillside on Santa Rosa Island and organized a field conference to study the remains in 1960. Later carbon dating showed the bones to be around 11,000 years old. Courtesy of the Santa Barbara Museum of Natural History.

Over the next days the little submersible had continuing problems coping with the currents. Again and again, just when he was getting close to a target of interest, MURV would be swept away, out of control. "The tides are blowing us out," Josenhans muttered one time. "We're getting so close," said Don. "Man, that's the nature of this game, frustration. We're just being dragged around." Another dive had to be aborted.

In the end, Josenhans got none of the cliff face images that he had hoped for. Don was especially apologetic about the poor performance of the sonar. It did not have enough range to help him keep track of where the submersible was relative to features on the bottom. As for using MURV to search for archaeological evidence, Don admitted what was obvious after watching the video images for hours: "You can't find artifacts with a ROV."

I was beginning to see just how limited the technical options were for undersea archaeology. Submersibles were difficult to operate and were unlikely to spot anything as small as a stone artifact, even if it were lying exposed on the sea bottom, which was unlikely. Divers might be able to dig and find artifacts in waters forty or fifty feet deep, but only with a large commitment of time and money. And in the Charlottes, with their history of incredibly rapid sea level rise, that would only take the story back a few hundred years beyond what could be found on the beach. Using the big grab jaws from a ship was the only practical way of probing the deeper and more ancient sea bottom off B.C.'s north coast.

Fedje's first attempts came during a few quiet hours in the *Vector* cruise. Josenhans was unwilling to commit much of the ship's limited time to dredging for artifacts, nor did Fedje expect him to do so. It was much too long a shot with the current state of knowledge. The cruise was devoted mainly to expanding their understanding of the sea level history of the area and completing their surveys. That would, they hoped, give future efforts at underwater archaeology a much higher likelihood of success. In the meantime, though, there were a couple of very obvious places where Fedje and McSporran thought it might be worth trying to bring up artifacts with the grab. These were in shallow waters just off two of their best archaeological sites, Richardson Island and Matheson Inlet.

Richardson came first. It was after dinner on the third evening in Gwaii Haanas waters, and the day's surveying had been completed. Fedje and McSporran finally got their chance. The ship hovered just off the beach where the water was about eighty feet deep. The deck crew rigged the big yellow dredging jaws from the large A-frame on the stern, with its steel cable running to a massive deck winch. They crossed their fingers and sent it down to take a big bite of the sea bottom. The bucket came up empty on the first two attempts. It had failed to close properly. The deck crew applied grease and hammered at the jaws to loosen them up. But finally the third attempt worked, bringing up half a cubic yard of dripping, oozing mud, gravel, and shell. Josenhans and Conway swung the bucket in over the aft deck and dumped the contents onto a couple of sheets of plywood. Then Josenhans and Fedje hunched over the soggy pile, poked at the visible stuff to see if there were any obvious artifacts, and made notes on the type of bottom material they were sampling.

Meanwhile, McSporran had affixed a wooden frame with a fine mesh screen outboard from the side of the ship, and hooked up a powerful fire hose. She put on high, bib-style rain pants to keep herself dry. The rest of us also donned rain pants and rubber boots. We shoveled the pile of muck into buckets and plastic boxes, labeled them, and carried them over to McSporran's screening area. This cleared the plywood for the next bucket to be brought up and dumped.

McSporran blasted the fine mud and sediments out of each screenful of bottom dredgings. Then she and Fedje picked carefully through the small bits that were left, hunting for the elusive piece of shaped rock that might turn out to be an artifact. "What's this?" McSporran asked a couple of times, handing a piece of sharp stone to Fedje. But he just knitted his brow and said he didn't think it was anything. One or two pieces were possibly worth a closer inspection, though, so he put them into little plastic bags.

The fourth bucket came up. And then a fifth. It was dirty and tiring work, shoveling and hauling the mud. Gradually, it got dark. The deck crew were tired, so they packed it in for the night. But there were bright lights on deck, so McSporran and Fedje kept working to screen the backlog of bottom dredgings. In the end, though, nothing of real interest turned up.

They repeated the procedure a few days later in Matheson Inlet, just offshore from the Arrow Creek site where Fedje had found the stone tool with a barnacle attached that dated to 9200 years. But the dredging jaws did not seem to be getting down deep enough to hit ancient deposits. "It's all turbated," said Josenhans, looking at that day's first pile of ooze on the plywood and rubbing it in his fingers. Worms living in the sea bottom had churned it up, meaning that a thick layer of recently deposited material was covering the seafloor. "It's all modern stuff," he scoffed. "Turbated mud. *Expensive* mud." They screened it anyway. The second bucket was the same. "Oh well, looks like it was homogenized by the bugs or worms," Josenhans sighed. The third bucket was similar. "Not good enough," Fedje scowled. And that was it for the dredging on that cruise. A total write-off. No banana.

And then time ran out. The scientists looked back on the effort as we cruised north to Queen Charlotte City. Breaking out a well-stashed bottle of wine, they shared it out in plastic cups on the foredeck. Fedje and McSporran agreed that dredging for artifacts really was like looking for the needle in a haystack. "Getting it with a bucket is pretty hit or miss," admitted McSporran. At least it would be until they had a much better way to narrow their search and target the likely ancient habitation sites. That's what the detailed surveying and coring were aimed at. "Even off Richardson, it was a shot in the dark," Fedje agreed. On the other hand, since it was so near a major archaeological site on shore, he still thought taking a few buckets was worth the effort. And the technique had long-term potential. "The bucket sampling method certainly could be useful for archaeology," said Fedje, "once we get a high resolution picture of the seafloor." This would be an exact digital model of such underwater features as drowned beaches that may have been stable for decades or centuries. These were the places where he could expect to find a concentration of artifacts. Then he would be able to select spots with lag surfaces, where the currents had winnowed the surface deposits so that the overburden mantling the bottom was not too thick to get down through. Finally, he could use the jaws to bring up samples from those sites.

All this preliminary information could only come from completing Josenhans' marine surveys. Even then, the chances would be very poor until the team had access to more advanced imaging technologies. Fortunately, such technologies were on the horizon. Obtaining much better images of the sea bottom and then targeting likely ancient habitation spots with the grab was the strategy they agreed to pursue. But it would take a few years.

CHAPTER TEN

Long Chronology

IT WAS THE sort of eureka experience an archaeologist might get to enjoy only once or twice in an entire career. Ruth Gruhn of the University of Alberta was working at a dig called Taima-Taima in Venezuela, at an ancient spring in an otherwise arid coastal region. It was 1976 and the remains of many extinct Pleistocene animals had already been found there in the 1960s: mastodons, extinct horses, and glyptodonts—huge creatures that resembled armadillos.

Gruhn was using her fingers to clear mud from the skeleton of a mastodon when she suddenly felt something hard. It turned out to be a fragment of a leaf-shaped stone projectile point, and to Gruhn its location told the whole story. It appeared to be lodged in the animal's pelvic cavity, which meant that it had most likely been used to kill the beast. As the dig progressed, she and her archaeologist husband, Alan Bryan, also found some sheared-off twigs, which they believed came from the animal's stomach. Radiocarbon dating showed the twigs to be around 13,000 years old. Gruhn and Bryan belonged to a minority of archaeologists who thought people first came to the Americas *well* before Clovis, almost certainly prior to the peak of the last glaciation, and probably 30,000 or more years ago. In other words, they endorsed a truly long chronology for people in the New World. Now they had their own site that apparently testified to pre-Clovis settlement in South America.

I was impressed by Gruhn's confidence about the Taima-Taima

dig and its interpretation. And since it was a coastal site, if valid, it supported the idea of pre-Clovis coastal migration. But other prominent archaeologists rejected Taima-Taima. Among them were leaders of a mainstream group that the long-chronology faction branded the "Clovis Mafia." C. Vance Haynes of the University of Arizona and Thomas Lynch (formerly at Cornell and more recently at the Brazos County Museum in Texas) looked closely at the claim made for pre-Clovis humans at Taima-Taima. They argued that most of the supposedly cultural items found there were not clearly of human manufacture, but could be naturally made objects and materials. The upwelling action of the spring, they said, had stirred up some of the sediments in the site. This cast doubt on the definitive association of the artifacts with the remains of the extinct animals. Finally, the underlying rock contained coal, which might have contaminated the organic samples with very old, so-called "dead carbon," and made the radiocarbon dates of the twigs come in older than they really were.

America's archaeological establishment set a high threshold of documentary proof before they would accept any site that might represent pre-Clovis occupation. They would only concur with such a claim if there were either datable human remains or undoubted artifacts found in an unquestionable, direct and undisturbed association with datable ancient organic materials, such as animal bones or charcoal. Moreover, the geology of the site had to show that the organic materials used for dating had not moved from one stratum to another or been contaminated by carbon of a different age. All archaeologists paid at least lip service to these standards of evidence. They would all love to work on digs where the facts were indisputable. In the real world, though, conditions at many sites were less than ideal. And then what?

There were those like Gruhn and Bryan who thought there were simply too many probable pre-Clovis sites in the Americas to ignore, and some were claimed to be much older than the 13,000 years of the Taima-Taima mastodon. Their interpretation of a broad range of evidence, especially from South America, was that people had been in the New World for 35,000 years or longer. Any particular site might fail to satisfy the very strictest criteria. Overall, though, there seemed to be so many likely older sites that they added up to a good case for pre-Clovis occupation. As Gruhn told me, dismissing her critics, those who refuted anything older than Clovis were becoming "desperate," because so many examples were appearing that contradicted them.

The Clovis Mafia adopted a directly opposite intellectual tack. To them, a lot of doubtful examples simply added up to an even more doubtful overall case. If there were so many examples of people in the New World long before Clovis, they asked their opponents, why can't you find even a single really solidly documented site? Why is there not even *one* like the early Folsom and Clovis sites, where the artifacts were clearly made by human hands and the associations with extinct animals or datable strata in the soil are direct and unquestionable?

As Thomas Lynch, arguably the *capo di tutti capi*, put it, "A number of archaeologists seem to believe that the marshaling of an increasing number of low-probability cases makes their argument for pre-Clovis man stronger. The opposite is true, both intuitively and statistically. The longer we look without finding anything certain, the more hopeless the pre-Clovis position appears—much as the multiplication of low probabilities yields yet lower probabilities. We need only one incontrovertible case, but we must have it to demonstrate the proposition."

And it was true that, over the years, whenever an archaeologist argued that a site dated to before Clovis, closer study showed significant problems with that interpretation. "So many have failed that the archaeological community has grown highly skeptical of any and all pre-Clovis claims," said David Meltzer, the archaeologist who was so dismissive of the idea that people could have skirted coastal glaciers in boats. He even marshaled some telling statistics. Taking a 1964 list of fifty allegedly pre-Clovis sites, he reviewed them and argued that by 1988 not one had survived scrutiny. "The shelf life of pre-Clovis claims," he quipped, "seems little more than a decade."

The Clovis First faction felt they occupied the high intellectual ground, but the controversy wouldn't die. Each disagreement over the validity of a site led to bitterness and hard feelings, in part because so much was often at stake for the archaeologist conducting a dig: professional prestige, bragging rights, future funding, promotion, and academic tenure. Moreover, archaeological projects seem to drag on forever. Funding delays can hold things up, and fieldwork can usually be done only during the summer season. In winter, there are courses to teach, conferences to attend, and papers to write. It is often ten or fifteen years between the time a promising site is first discovered and the time its analysis is complete and published. Many archaeologists only get to lead two or three major digs in their entire careers. Naturally, after devoting so many years of their lives to an effort, they have a strong interest in defending their results against criticism.

From the late 1970s through the mid-1980s, Tom Dillehay of the University of Kentucky excavated a site a day's march from the coast in south-central Chile. There was excellent preservation of organic materials and there were numerous stone tools, and radiocarbon dating had placed the main artifacts at around 12,500 to 13,000 years old. Like Taima-Taima, if valid, this age would overthrow the Clovis First scenario. Another section of the site had fractured stones that Dillehay interpreted as artifacts—the soil stratum there dated to around 33,000 years. Dillehay did not make a very strong claim for the validity of these much older possible artifacts. The 12,500 year date, since it was significantly older than Clovis, was controversial enough.

As with the Venezuela site, Monte Verde was initially rejected by some heavyweights of American archaeology. The first was Junius Bird of the American Museum of Natural History, who in the late 1930s had excavated Fell's Cave in Patagonia. This was the site, radiocarbon dated to between 10,800 and 11,000 years, that forced archaeologists to envisage a very rapid colonization of the hemisphere right down to the tip of South America.

Bird first visited Monte Verde during a preparatory stage very early in the excavation. His schedule was apparently determined as much by personal considerations as professional ones; he was en route to meet friends for a yachting cruise. According to Dillehay, Bird initially stayed only about forty minutes, mainly telling stories about his own exploits. And then he applied preconceived notions of what a Pleistocene site in South America *must* contain. When he looked at the artifacts found up to that time, he asked only, "where are the projectile points?" Without such artifacts, he would not give credence to Monte Verde as a Pleistocene site. Bird was supposed to return for a closer look once the excavation proper had begun, but instead he arrived a week earlier than planned, while Dillehay and his crew were still removing the overburden. "He stayed two nights and two days and left," Dillehay later grumbled to a journalist. "He didn't see the excavation of artifacts *in situ.* If he'd come one week later, he would have seen projectile points associated with a mastodon."

Word filtered back to Dillehay that Bird had gone off and told colleagues in the capital, Santiago, that Monte Verde was not an archaeological site at all. He also allegedly wrote a letter denigrating the site to the National Geographic Society, which had been supporting Dillehay. The Society cut off funding a year later, and it proved very difficult for Dillehay to find money elsewhere. Dillehay was embittered, but he did manage to continue his excavations. When he began publishing his results, Monte Verde was quickly recognized as a likely pre-Clovis site. In his book, *The Settlement of the Americas*, Dillehay accused the now-deceased Bird of grossly misrepresenting the archaeological record in South America.

Not all allegedly pre-Clovis sites were from South America. The most widely accepted one in North America was a limestone rock shelter at Meadowcroft, Pennsylvania. It was excavated beginning in the mid-1970s by a team led by James Adovasio of Mercyhurst College. Hundreds of stone artifacts were found there in eleven different strata. These were extensively carbon dated and found to represent human occupation right up into recent prehistoric times.

Controversy arose, though, regarding the strata that appeared to date to between 13,000 and around 19,000 years in one case and between 11,000 and 13,000 in another. The problem was not so much with the artifacts, which undoubtedly reflected human presence, but with the dating. Critics argued that the older levels contained no telltale extinct Ice Age mammals, which made them suspicious. Because some charcoal found at the site was about 31,000 years old, the artifact-bearing strata might have been contaminated by coal or other older organic matter.

After two decades of replying to detailed critiques and contending with intense scrutiny, Adovasio and his colleagues have largely recanted their "previous and probably overexuberant pronouncements" for the oldest stratum. They now only claim certainty that the site was initially occupied by around 12,000 to 12,800 years ago. The true age of Meadowcroft remains in question and may never be resolved.

The most contentious site in South America was a rock shelter called Pedra Furada, a deep indentation at the base of a cliff some 500 miles inland from the coast of northeastern Brazil. It was excavated by French-Brazilian archaeologist Niéde Guidon. She found hundreds of stone flakes and other fractured stone pebbles that she claimed were artifacts. They were in stratigraphic layers that were dated using charcoal from what she interpreted as campfires in hearths. Those dates, which ranged initially between 17,000 and 32,000 years, raised the hackles of the Clovis First mainstream because they implied such a long chronology. It was just the beginning. Guidon went on to excavate other sites in the region and reported that some Brazilian rock shelters were occupied at least 45,000 years ago, if not earlier.

A number of archaeologists countered that what Guidon had found were not confirmed artifacts at all, but more likely "geofacts," stones that had been chipped and flaked naturally by falling from the high cliff. They thought the ancient hearths were probably just the remains of naturally caused fires. Three of the most prominent critics were North Americans who attended a conference in Brazil and then inspected the site in 1993 and discussed it with Guidon. One was David Meltzer, the others were James Adovasio and Tom Dillehay, whose own putative pre-Clovis sites had been subjected to such close scrutiny and criticism.

Meltzer's later comments reflect his attachment to the Clovis First model. "We are concerned that the hearths are in fact hearths and not natural fires. Are the pebbles truly artifacts—or has nature been mischievous? In the last thirty years there have been thousands of pretenders to the pre-Clovis crown. Each one has dissolved under scrutiny. So we have got pretty skeptical. . . . If we have pre-Clovis humans in South America, then by golly, why don't we have them in North America too?"

Neither Adovasio nor Dillehay could be accused of lack of sympathy to the idea of pre-Clovis people in the New World. But they joined Meltzer in coauthoring a detailed critique of Guidon's conclusions in the journal *Antiquity.* It outlined their "concerns" and offered suggestions as to how she and her colleagues could bolster their case. Until then, they concluded, they could not "accept" Pedra Furada as a Pleistocene site.

Guidon's rejoinder was bitter, in part because she had agreed to the organized inspection of her site and felt her U.S. peers had acted in bad faith. The "battery of questions" they subsequently raised surprised her, she wrote, because "none of the three colleagues came up with these questions during the 1993 meeting—mounted precisely to generate direct dialogue on the peopling of the Americas." She implied that they lacked "scientific honesty" and the necessary expertise on archaeology in tropical regions, and said their "commentaries are worthless because they are based on partial and incorrect knowledge."

In an atmosphere seething with rivalry and distrust, questions of scientific fact even turned political. In 2000, Guidon gave an interview on her work to Britain's left-leaning daily, the *Guardian.* "I don't have any doubt that the oldest traces of humans yet discovered are here in Brazil," she said. By then she was arguing that the dates might go back as far as 50,000 years. The *Guardian* even presented

Guidon's work as a potential "blow against U.S. imperialism." Nor did Guidon's comments dispel the overtones of nationalism. As she told the journalist, "The problem is that the Americans criticize without knowing. The problem is not mine. The problem is theirs. Americans should excavate more and write less."

Had the First Americans arrived 12,000 years ago, or was it 50,000? Disagreement among archaeologists was an open invitation for other sciences to weigh into the fray. Beginning in the early 1980s, a whole series of highly creative, but indirect approaches were taken that purported to show who the First Americans were, along with when and how they came to the New World. Some were based on genetics, others on linguistics or on measurable physical characteristics, such as tooth patterns or skull shape. Archaeologists generally had little faith in the conclusions, but they certainly captured public attention. Many of these alternative approaches supported a long chronology.

The first, and most widely publicized in the mainsteam media, was the so-called "three migrations theory." Until then, the prevailing view was that the New World had seen two major population influxes. The first (the overland hunters coming across the Bering land bridge around 12,000 years ago) were assumed to be the ancestors of all the American Indians on both continents. A much later migration, perhaps only 4000 or 5000 years ago, had brought the "Eskimos," whose appearance and culture seemed so different from that of "Indians."

The American Indians, though, were an extremely diverse population, and pre-Columbian America was a virtual tower of Babel. Comparative linguist Joseph Greenberg of Stanford University spent decades filling twenty-three notebooks with lists of words from the roughly 1000 aboriginal languages that existed when Europeans arrived. Most linguists claimed that these languages could be grouped according to common traits into anywhere from 100 to 200 language families. By definition, each family of languages was distinct enough from every other that no features of grammar or vocabulary allowed them to be "lumped" together into larger groupings. Greenberg stirred up an academic hornet's nest when he claimed to discern commonalities that let him group all these tongues into three linguistic "superfamilies."

By far the largest was what Greenberg called Amerind, comprising all the aboriginal languages of South and Central America, and most of the languages of North America as well. The second major group was Na-Dene. It comprised the languages spoken by the Indians of the continent's far Northwest, the largest being Athabaskan, but also including the coastal languages of the Haida and Tlingit. The third group was Eskimo-Aleut. It comprised the languages of the Aleutian Islanders, along with those of Alaska's Yuit and the much more widely spread Inuit, the latter stretching from Alaska right across the far North to Greenland.

Greenberg suggested that each superfamily had been implanted by a discrete ancient migration of people from Asia into the Americas. In each case, those first migrants, probably few in number, would have brought along a single language or closely related group of languages. Once they arrived, expanded in population, and spread out in America, they would have lost contact with each other. Their languages would have branched off, like the limbs of a tree, and grown apart. But the original trunk, or "proto-language," might still be dimly perceived, Greenberg thought, because of a very small number of words and word fragments that many of the languages still had in common. In Amerind, for example, Greenberg identified a tendency for the letter "N" to mark the first person pronoun and "M" the second person.

Greenberg's three-way split was supported by dental anthropologist Christy Turner II of Arizona State University. He had combed the storerooms of museums around the world, measuring and studying jaws and teeth from hundreds of long-dead people. He also studied countless living ones. The shape of a person's incisors, the number of roots for certain molars, and the number of cusps on the crowns of molars all reflect heredity. Using pre-Columbian teeth helped Turner to avoid the complications posed by genetic mixing since the arrival of Europeans and Africans in the New World.

Turner found that the overall configuration of nearly all Native American teeth was similar to that of ancient peoples from northeastern Asia, as opposed to the teeth of southeastern Asians or Polynesians. Finer distinctions allowed him to divide them into three groups that corresponded to Greenberg's linguistic classification: the Eskimo-Aleuts, the Amerind-speaking peoples of most of North and South America, and the Northwest Indians who speak Na-Dene languages. He argued for three separate migrations from geographically distinct parts of Asia.

There were few other dental anthropologists to critique Turner, but many academic linguists greeted Greenberg's analysis with howls of outrage. One argued that it was so speculative that it should be "shouted down." Another said that reading Greenberg's "principles" of language classification made him feel "more and more like Alice in Wonderland." A third called Greenberg's techniques "so flawed that the equations it generates do not require any historical explanation, and his data are unreliable as a basis for any further work." And an anthropologist with extensive Alaskan experience thought the division of Native Americans into three groups was just a little too neat to be true. He quipped that "researchers who flirt with trinities should be reminded that Eskimos have walked on water for 10,000 years. They wait for it to freeze, and when on this ice, they avoid making unnecessary waves."

Greenberg and Turner basically accepted the Clovis First time frame, which implied that the first migration, by the Amerinds, must have reached the middle latitudes of North America after 12,000 years ago. The other migrations would have come later. Evidence from a number of different geneticists also seemed to support their fundamental division of Native Americans into three groups. But some of that evidence implied a much longer chronology than allowed by Clovis First.

Canadian physical anthropologist Emöke Szathmary (today president of the University of Manitoba) collected blood samples from native groups across the North. She studied genetic "markers" to measure relationships among the groups, which might indicate ancient kinship and origins. Human proteins, enzymes, and blood group antigens are determined by genes. Similarities between two individuals in specific proteins or antigens indicate potential genetic relationship. Long before DNA testing existed, these similarities were commonly used by courts to determine paternity. Such genetic traits are distributed in very different frequencies among different populations, and the distributions can be studied statistically to infer different evolutionary paths. By carefully correcting for recent "admixture" (the infusion of European and African genes in historic times), Szathmary set out to unravel the relationship between the ancestors of Indians and Eskimos in ancient times.

It was a tricky exercise, with inadequate samples and considerable overlap in the distribution of traits, but the results were intriguing. Like Greenberg and Turner, she found three groups. The large group of northwestern Indians who speak Athabaskan and related languages were (unexpectedly) much closer genetically to Eskimos than were the other Indians of North America. This was even true where the "other" Indians, such as the Algonkians of eastern Canada, have lived in geographical proximity to Eskimo groups during recent millennia. In fact, Athabaskans from northwestern Canada were closer genetically to the far-off Greenland Eskimos than they were to geographically proximate "other" Indians of the Canadian Northwest.

To explain this, Szathmary concluded that American Indians and Eskimos might derive from two peoples that were already distinct in Asia, likely as a result of millennia of evolution in geographically separated territories, before their descendants moved separately into Beringia. But subsequently, the Indians could have become divided by the great Canadian ice sheets into northern and southern groups, and those groups proceeded to evolve separately for many thousands of years.

In her scenario, then, Athabaskan Indians evolved from ancestors who had remained north of the ice, while all other Indians evolved from ancestors who were already south of the ice well before the Ice Age ended. Over the thousands of years of separation, the northern group must have mixed to a genetically recognizable extent with the ancestors of the Eskimos, which implied that the latter, too, were already in Beringia and were not relatively recent migrants. (Assuming that the ancestors of modern Eskimos were already a coastal people, the archaeological record might be silent about them due to the subsequent rise in sea level.)

Then, as the ice sheets melted, the ancestral Athabaskans spread from Beringia across much of Canada's northwestern interior, and to the coast of Alaska and British Columbia as well. (The Tlingit of coastal Alaska speak a related language that clusters with Athabaskan in the Na-Dene grouping. Some linguists consider Haida to be a distantly related Na-Dene language, as well. Others believe that Haida has no recognizable connection to any other known language; they call it an "isolate.") But if people were also south of the ice sheets and cut off from Beringia for thousands of years, this would put them there before the peak of glaciation. Because of these findings, Szathmary was inclined to accept James Adovasio's initial claims for his site at Meadowcroft in Pennsylvania, with its hint of occupation as long as 13,000 to 19,000 years ago. Hers was the first evidence from genetics that tended to support a relatively long chronology. Much longer chronologies were to come.

A large group of geneticists has attempted to understand the pattern and timing of migration into the Americas by looking at changes over time in mitochondrial DNA (mtDNA). Unlike the better known "nuclear" DNA found in a human cell's nucleus, mtDNA is found in tiny "organelles" (the mitochondria, which control energy production) outside the nucleus. For science, mtDNA has the advantage that it is passed down only along the maternal line, which makes it simpler to trace back. It also mutates much more rapidly, which makes it useful as a "genetic clock." MtDNA first made headlines in the 1980s when a team at the University of California at Berkeley, announced that they had traced all human family trees back to a single woman in Africa from whom the mtDNA of all living humans is supposedly descended. By applying the average rate at which mutations accumulate in mtDNA, they estimated that this woman, inevitably dubbed "Eve," had lived around 200,000 years ago.

In the 1990s, a long series of studies showed that the mtDNA of Native Americans included four readily identified lineages, which are distinctive sequences of genetic code that first appeared through mutations and were subsequently passed down along the maternal line. Each lineage (labeled A, B, C and D) was also widely found in mtDNA from Asian peoples, but not elsewhere. (By comparison, 99 percent of European mtDNA includes nine *other* lineages.) This pointed toward migration from Asia and came as no surprise.

But the four lineages were unevenly distributed among Native American populations. Several studies showed that Amerind-speaking peoples had all four in substantial frequencies, while among the Na-Dene A was predominant and among the Aleuts and Eskimos A and D were most common. The extent of genetic diversity in mtDNA could also be used to construct phylogenetic "trees" (branching diagrams or "dendrograms" that look a bit like family trees) purporting to show approximately how long ago the lineages diverged and how closely or remotely today's living peoples were related to each other. A number of scientific teams also looked at the uneven distribution of these mtDNA lineages in Asia and tried to determine from such data the specific geographic regions from which the ancestors of different Native American groups may have migrated.

The results have been ambiguous and contradictory. Some geneticists have concluded that the "homeland" of nearly all Native Americans is Mongolia. Others trace different Native American groups back to several geographically distinct regions in Asia. Some argue for three or more discrete migrations from Asia at different times. Others perceive a single wave of migration from Asia into the New World. The research and debate continues.

Almost the only consensus has been widespread agreement that some (and possibly all) of the mtDNA lineages date back anywhere from about 14,000 to 38,000 years. But this is a pretty broad range, and the exact "time depth" resulting from each study depends on the assumed rate of mutation in mtDNA, which cannot be independently confirmed for the New World lineages.

Throwing a curve into the entire question of how many ancient migrations there may have been is the existence of a fifth, and much rarer, lineage, designated "X," which has turned up in a very low percentage of Native Americans. Elsewhere it is centered in Europe and the Middle East. Scientists still don't know how or when it came to the New World. Did it spread at a very low frequency across to Asia and reach the Americas that way? This mysterious genetic influence has engendered much speculation. Some think there is a link to the supposedly "Caucasoid" features of the Kennewick Man skull. Most of the scientific spinoff has been a legitimate attempt to explain the facts as they are known at this time. Perhaps there was a fourth ancient migration, across the Atlantic?

These "European" genes have also sowed some nasty weeds in the fertile intellectual garden of voodoo anthropology, where virulently racist authors have crawled out from under the rocks. One example is a tome called *March of the Titans: A History of the White Race*, distributed over the Internet. A chapter on "Lost White Migrations" looks at supposedly Caucasoid remains like Kennewick Man and also at the anomalous "Lineage X—The White Link Shown by Genetic Tracking." Kennewick Man was found with a spear point in his pelvis, and lineage X may have largely died out in America. This leads the (virtual) book to conclude that "the first Whites in America were killed in open warfare with Amerinds (who may have arrived simultaneously, or afterwards) and that the survivors were absorbed into what became the numerically dominant Amerind groupings. The existence of the Lineage X gene string adds credence to this. As mtDNA is transmitted only through the female line, it is obvious that the White males were killed by Nonwhites, and the White females were taken alive by the Amerinds for sexual purposes." This venom is being spewed across cyberspace.

* * *

As with genetics, some linguists have tried to use language evolution as a ticking clock to estimate how long people have been in the New World. Languages evolve over time much like biological species, diversifying, branching off like the limbs of a tree, and becoming increasingly distinct. Johanna Nichols of the University of California at Berkeley looked at the linguistic diversity of pre-Columbian America. Finding that the Americas had about the same amount of linguistic diversity as Australia and New Guinea, which have been inhabited for 40,000 to 50,000 years, she began to suspect that the New World must, therefore, have been populated for about as long as those territories. In any case, she was certain that the amount of linguistic diversity in the Americas could not be accounted for in the 12,000 or so years allowed by Clovis First.

Then she spent years crunching numbers. She found about 150 language families in the Americas. Looking at the rate of language divergence in places like New Guinea, where the time of first occupation is relatively well determined, she found that each language family could be expected to engender one and one-half to two new language families about every 6000 years. Assuming two or three waves of migration, she factored in this rate of language evolution and concluded that the first wave or two of migrants likely arrived from Asia 30,000 to 50,000 years ago.

As with conclusions based on estimated rates of genetic mutation, though, there was no control on this experiment, no way to verify that the rate of language diversification in the Americas has actually followed the pattern set elsewhere. If for any reason the assumed rate was off by a significant factor, all the calculations were out the window. The long chronology remained hypothetical.

A very different type of linguistic argument was made by Richard A. Rogers, a geographer formerly at the universities of Nebraska and Kansas and later at the U.S. Department of Agriculture. Personally, I found it far more compelling.

A basic principle of linguistics asserts that the diversity of languages in an area reflects the length of time those languages have been there. As Rogers told one academic symposium, "as languages diversify, the longer they were in a geographical area, the more diverse they would become." If you look at the geographic range covered by a group of related languages, therefore, "the area that shows the greatest diversity has a very high probability of being their homeland." What's more, as the languages diverge further, language groups arise that do not seem to be related to other language groups in the vicinity. This means that any region with a tight clustering of many seemingly unrelated languages is probably an area that has been populated for a very long time. By contrast, any geographic region that has only one or a few languages, and where they are spread out over a large area, has likely been populated relatively recently.

There was no denying his highly original insight. The geographic clustering was clear and startling, I thought, and unlike some of the other theories, it did not depend on the tricky business of estimating rates of diversification in language or genetics.

Rogers drew up maps of North America with dots representing the probable linguistic homelands at the time of the first European settlers. Not only were these points very unevenly distributed, most were outside the area that was once covered by glacial ice. Equally striking was the concentration of language homelands on the coasts, especially the Pacific.

"Look at that map," he said, with its concentration of points along the North Pacific coast from Alaska to California. The coastal area was home to many very distinct tongues in close physical proximity. Haida, for example, is very different from the adjacent Tlingit language. Farther south, there were other clusters of very diverse languages along the Pacific coast of Central America and the northern coast of South America. By contrast, the interiors of both continents had few distinct languages, and most of them extended over very large areas. To Rogers, this implied that people had settled the coastlines first and spread only much later to interior areas.

Ruth Gruhn took up Rogers' argument and elaborated it further. She looked at languages along the Gulf of Mexico and on the Atlantic coast of South America as well. There were many more diverse Native American languages on the U.S. Gulf Coast and in Florida, for example, than in the interior of the southeastern United States.

Gruhn proposed a "coastal entry model" for the first peopling of the Americas. It had the migrants first coming into the hemisphere along the North Pacific coast near the Bering Strait in a slowly spreading and narrow carpet of coastal occupation. They would have followed the Pacific coast south as far as Panama, then some would have crossed the narrow isthmus and followed the shores of the Gulf of Mexico, both northward, eventually bringing them to the southern coast of the United States, and eastward along the northern coast of South America. Meanwhile, others would have carried on southward, following the Pacific coast of South America.

In agreement with her husband Alan Bryan, Gruhn strongly endorsed a long chronology. She believed that a number of sites, especially in South America, were 30,000 and more years old. The only way to account both for those sites and the coastal linguistic concentration was to posit a single, very early coastal migration starting as much as 50,000 years ago at the Bering Strait. But if the claims for very early occupation in South America were wrong, nothing inherent to the coastal entry model required such a long chronology. The migration along the coast from northeastern Asia could have started "only" 15,000 years ago, for example, and still account for the linguistic facts as Gruhn outlined them.

By the mid-to-late 1990s, many threads of nonarchaeological evidence ran counter to Clovis First. They were provocative approaches, and some of them provided encouragement for the proponents of coastal migration. But such theories by no means demonstrated that people had actually been at a particular place and time. Only archaeology could do that.

CHAPTER ELEVEN

Maverick Archaeologist

I WOKE TO unfamiliar sounds and needed a moment to figure out where I was. Someone was moving around in the drying tent that stood next to mine, between the beach and the steep wooded hillside. It was the start of my second day on Prince of Wales Island. The alarm clock said six A.M., but it gets light by around four on the Alaska coast in August, and Tim Heaton was already up, screening some of the cave diggings from the previous day.

Hoping to grab a few final winks, I wriggled back into my sleeping bag, but soon the rest of Heaton's young paleontology crew were bustling around, too, getting ready to hike up to the cave site for breakfast. Actual work on the dig started around eight, and I didn't want to miss the morning meal or any of the research action, so I assembled my gear and trudged up that killer trail. By the time I pulled myself up the last rope and rock face and approached the clearing, I was sweating. I wiped my brow, straightened my clothes to look presentable, and plodded up the last stretch.

A couple of young guys I didn't recognize were eating breakfast in sunshine on the steps that led to the cooking yurt. They seemed to be nursing hangovers, and I dimly recalled the night before. The archaeology crew had taken the previous day off, piled into the team's big Zodiac, and motored off to what passed for civilization. After doing laundry and indulging in showers, they had hit the floating bar at Point Baker for a bit of a blowout. Like Port Protection, where I had arrived by floatplane, Point Baker was a roadless community with about fifty people, who mainly trolled for fish and lived on boats or in ramshackle houses shoehorned into a minute harbor. By the faint light of late summer dusk, the archaeologists had made it back to the mooring just off the beach campsite. Most of us were already in the sack. Staggering past our tents, they had found their way up the dark trail using flashlights.

Jim Dixon—I was meeting him for the first time—was sitting just outside the yurt. He was in his early fifties, clean-shaven and a bit chunky, with a round face that lent itself to puckish, amused expressions. When I showed up, he was busy proofreading the galleys for his latest book. I ducked into the yurt to pour myself a cup of coffee and pile some pancakes on my plate. After breakfast, Dixon suggested that we find a quiet spot to talk down near the mouth of the cave. We sprawled out on the dry forest floor next to his checkerboard of partially excavated squares.

As Dixon told me with barely concealed pride, he had always been something of a maverick among American archaeologists. Although raised in New Jersey, he'd had a lifelong connection to Alaska and its coast. His grandfather owned land on the Kenai Peninsula, and Dixon had spent some idyllic summers there, falling in love with a shoreline where ice-capped mountains plunged into the sea, where the mist hung low over the fjords, and you could watch a grizzly swipe a salmon from the spawning stream. He was supposed to be helping his grandfather build a cabin on the property, but mainly they spent time together out on the water, fishing. He liked it so much that he ended up doing his undergraduate studies at the University of Alaska.

Dixon left Alaska to get his Ph.D. from Brown University, but returned for his doctoral research. He had accepted the prevailing idea that the first people in the Americas had walked across the Bering land bridge, and his dissertation involved trying to locate those earliest archaeological sites, both inland and on the coast. "I actually did a marine archaeological survey off the Pribilof Islands," he told me. The Pribilofs lie well out in the Bering Sea today, but when sea level was low they would have been on the southern shore of the land bridge. Although Dixon did not yet imagine people traveling all the way from Asia by boat, he realized that coastal areas offered a wealth of resources even to land-based people. Now, decades later, he savored the memory of himself as a brash young graduate student, "looking for archaeological evidence of people living on the margins" of that long-drowned territory. "That was in 1976. As far as I know, that was the earliest such survey done. When I did that work, it was highly criticized. People said, 'Well, if you can't find this stuff on land, how do you expect to go out there and find it in the ocean?' "

At the time of my Alaska visit, Dixon was director of the Denver Museum of Natural History, although he soon moved to a professorship at the University of Colorado in Boulder. But his first academic posting had been at the University of Alaska in Fairbanks, both as a professor and curator of the university's museum. His early fieldwork was largely spent excavating caves on the Porcupine River in northeastern Alaska. These were in an area that had been ice-free during the last glaciation and therefore might well hold traces of very early people. If the Clovis First scenario were correct, he should find artifacts or other human indicators well over 11,000 years old.

He found, instead, that nearly all early artifacts (to the extent that they could be solidly dated) fell into the 8000 to 10,000 year-old range. This made the region inconclusive as to the first peopling of the Americas. But the work gave Dixon valuable experience with the problems and pitfalls of cave archaeology, including the danger of assuming that he had found a legitimate artifact when in fact the situation was doubtful. In caves, he said, there were lots of stones and bones that had been broken by natural processes. Rocks, for example, could fall from high ledges and shatter, leaving the kinds of sharp edges typical of artifacts. Bones could be trampled and broken by huge Ice Age animals, inspiring similar misinterpretations. Both situations were known to archaeologists from ancient caves in places and times where people could not have been present. He also became an expert on the prehistoric tool traditions of Beringia.

Dixon initially accepted the established view that Clovis big game hunters were the first people to reach mid-latitude North America. If so, then the oldest archaeological sites in the New World should be located in the ice-free interior of Alaska and "it was obviously just a matter of time until they were discovered." Since Clovis points were characterized by their distinctive fluting, he expected that fluted points older than Clovis would eventually turn up in central or eastern Alaska, near the northern end of the "ice-free corridor." When they did not, even after many years of searching, he began to have niggling doubts.

Some stone tools that slightly predate Clovis *were* eventually found in central Alaska, but not until the mid- to late 1980s, when his thinking was already evolving. These implements came from a few sites with a tool complex that archaeologists call Nenana. The very earliest artifact-bearing strata in these sites date back 11,700 or 11,800 years. Most Nenana sites, though, are from the same time period as the Clovis sites a couple of thousand miles to the south. The new book Dixon was proofreading attempted to explain the connections, if any, between Nenana, Clovis, and other northern tool traditions.

The Clovis and Nenana complexes, he told me, "were mainly contemporaneous, and both seem to have appeared without precursors in very different parts of North America at almost exactly the same time." Along with other stone tools that were used for various purposes, the Nenana artifacts included triangular and teardrop-shaped projectile points. These lacked the telltale Clovis fluting for attaching a shaft, and they were considerably smaller than Clovis points as well. To Dixon, therefore, Nenana did not look like a precursor of Clovis. What complicated the picture was the fact that, over decades of searching, some fluted points *had* in fact been found in eastern Beringia. However, none of them had been unearthed in archaeological contexts that allowed them to be firmly dated. Many of these undated points had found their way to museum collections in Ottawa, New York, and Washington, D.C., and Dixon inspected those artifacts.

In mid-latitude North America, the Clovis style of very long fluted spear points had been followed by the Folsom tradition of considerably smaller points with a somewhat different shape and design. Timewise, Folsom overlapped Clovis somewhat and continued until around 10,000 years ago. To Dixon, the Alaska fluted points he examined looked much more like Folsom than Clovis. If he was right, the toolmaking influence, and possibly also the people practicing that technology, had moved northward instead of southward.

"I published an article suggesting that this Paleo-Indian tradition, these little points, actually went from south to north." This was mainly based on Dixon's sense that "the ones in the north looked typologically younger in the spectrum." He also took into account the limited geological evidence available on the ice-free corridor. It was inconclusive, but Dixon did not think the corridor had opened in time for Clovis to have come down from the North. His contrarian scenario was that "the corridor opened, and then people or traditions spread from south to north as the glaciers melted around 11,000 to 11,500 years ago." A few prominent anthropologists, notably James Griffin of the University of Michigan, agreed.

This still left the question of how Clovis people reached temperate North America before the corridor opened. Dixon's experience on the Alaska coast had left him open to the idea of early coastal migration. He recognized, of course, that the evidence would be very hard to find because of the postglacial rise in world sea level. On the other hand, the notion of a broad coastal zone along the fringes of Beringia—one that was now drowned but could once have been inhabited—provided a missing link in the chain of archaeological logic. Tool traditions that predated *both* Nenana and Clovis could have existed there. These traditions could have moved south along the coast as far as the unglaciated latitudes well before 12,000 years ago. Perhaps as they evolved they only moved inland later, leaving Nenana tools in the north and Clovis points in the south.

The final major influence on Dixon came in 1987 when Tom Dillehay came to Alaska to give a talk. Dixon went to greet him at the airport and was impressed by his neat, even fashionable dress and appearance, in contrast to the scruffiness of so many Arctic archaeologists. Dillehay had spent half his adult life in South America and spoke excellent Spanish. He had already published some of his key findings from Monte Verde and reported that the main part of the site dated back around 12,500 to 13,000 years. If correct, Monte Verde was 1000 years older than either Nenana or Clovis.

Dixon never inspected Monte Verde for himself. "But Dillehay's work did impress me a lot," he said. "When I first heard him talk and saw his presentations, I could tell that he was very competent. And the data looked very good to me. I mean, it just looked solid." The main problem with it, he laughed, and one reason why it took a decade or more for many others in the archaeological world to accept it, "was that it didn't look like anything that we had predicted we would find. It wasn't a big-game hunting site or anything like that," but one where people had engaged in more general foraging for a wide variety of foods.

If he were to accept Dillehay's work, Dixon had to explain how people might have reached Monte Verde that early. Coastal migration was a possible solution. It ran counter to entrenched opinion, of course, but the more he thought about it the more sense it made.

Dixon knew that Australia and the Solomon Islands had been colonized well before the last glacial maximum. "Clearly watercraft have been around a long time. Notice that I'm not calling them 'boats,' " he laughed. They might just as well have been some kind of rafts. Nor did he doubt that people lived on the Bering land bridge, even though no evidence of them has turned up near the Bering Strait. "I don't think it's an either/or thing. But in terms of an entry route to the Americas, I think that people would have been blocked in eastern Beringia by the ice. Whereas they would have been able to come along that coastline earlier."

Dixon saw the coastal route as a much easier and safer one for sizable groups of migrating people. Children, pregnant women, and old people, he argued, could not have slogged their way very readily across the tundra. And they would all be dependent for food on the few adult, male big-game hunters in their band. On the coast, though, "shellfish beds can be harvested by children and the elderly, and watercraft enable family units to travel in much greater safety and numbers." Traveling along the coast, moreover, there would have been no need to make drastic adjustments to changing climate and food resources, no need to develop new ways of hunting, or make new kinds of tools or weapons. Their familiar techniques, tool kit, and life ways would have served them well along a migration of many thousands of miles and many degrees of latitude. "If you look at the resources in Kamchatka and northeast Asia, and along the North Pacific Coast, and you look at them on the Northwest Coast, they're very similar. You have salmon and shellfish and seals. So I think once you have that adaptation, moving along that coastline is very easy."

Finally, Dixon pointed out that many people who are unfamiliar with coastal Alaska imagine that the area is blasted by gale force Arctic winds nearly all the time. This would indeed constitute a great difficulty for coastal navigation in simple rafts, dugouts, or skin boats. And Ice Age winters might truly have been a living hell, with long months of darkness and howling blizzards. But people did not have to travel in those conditions. They only had to lay in food supplies, hunker down in caves or other shelters, and survive until summer. Just as on the day of my flight north, the summer weather remained mild and sunny and the sea extremely calm throughout my visit to Prince of Wales Island. A big North Pacific high-pressure system had moved into the Gulf of Alaska and was just sitting there. "Even in the Bering Sea you can get stretches of weather like this," he said. If people on the coast were patient, watched carefully and waited for such periods of stable good weather, travel could be safe and predictable.

As his ideas evolved, Dixon gave full credit to Knut Fladmark and joined him as one of the first few academic archaeologists seriously promoting coastal migration. In 1990 he completed a book, *Quest for the Origins of the First Americans.* Although it did not actually appear until 1993, *Quest* used the coastal migration scenario to explain the archaeological patterns he had discerned in Alaska. He wrote that both Clovis and Nenana may be "derived from a common technological tradition that predates both . . . and possibly represents the first humans to enter the Americas." To account for the pre-Clovis appearance of people at Monte Verde, he suggested that "prior to 11,500 [years ago] and possibly as early as 14,000–13,000 . . . people had spread south along the west coasts of North and South America."

When he showed his book manuscript to colleagues, Dixon discovered just how controversial such ideas could be. "Yeah," he chuckled. "I took some lumps for that." One very senior colleague who read the draft prior to publication told him, "you'd better kind of tone this down, you're really stretching your credibility." That was "not an establishment book," said Dixon, "but over time, you know, it's been almost prophetic. It's been kind of a foreshadowing of where we've come."

Dixon stuck his neck out even further. In the conclusion to the book he indulged in speculation to explain some of the more equivocal South American finds suggesting a *really* long chronology. There was Dillehay's lower level at Monte Verde, with what might be artifacts dating to 32,000 years ago. And there was Guidon's Pedra Furada site in Brazil, for which at that time she was claiming stone tools dating back 33,000 years. Yet there were no sites in North America with reasonably secure dating that were anywhere near as old. (Adovasio's Meadowcroft site in Pennsylvania was the oldest site that Dixon at the time thought just *might* be valid.)

In *Quest*, Dixon suggested that if those South American sites were valid, they most likely pointed to extremely long sea voyages from Southeast Asia or the southwestern Pacific islands to South America. In fact, he—personally—had strong doubts about any such ancient, long waterborne migration. He was much more inclined to believe in stepwise migration around the North Pacific and a relatively short chronology for the first settlement of the Americas, likely 15,000 years or less. When I got to read the galleys of his latest book and discuss his latest ideas, I understood better his reasoning for the shorter time frame. But he had no regrets about the opinion he had expressed a decade earlier. "I simply said we should keep our minds open on that," he insisted, raising an eyebrow as if to ask whether I was with him on this. "I still think so."

The excavation's crew chief, archaeology graduate student Craig Lee, interrupted us to confer with Dixon about one of the squares he and his four diggers were currently working just outside the mouth of the cave. There was a large rock in one corner of the square that they decided had to be sketched and then removed. Meanwhile, the paleontology crew were busily working inside the cave, and carrying their bags of diggings out for the preliminary wash in buckets of water.

The morning's work moved along uneventfully. The generator droned away, and the diggers dug. Then everything stopped. After breakfast, Heaton had gone back down to the beach and off in the Zodiac to fetch a local Prince of Wales resident named Pete Smith. Heaton needed some specialized help. Smith was a short, wiry guy who made his living running a logging and sawmilling operation. He was also a fellow caver who had been involved in the earliest exploration of the island's limestone caverns.

Heaton's excavation mainly involved removing and studying sediment. That's where the interesting stuff was to be found. But occasionally, large chunks of dumb old limestone also had to be removed. Recently, several slabs of rock had fallen from the ceiling and walls of the cave. It was impossible for Heaton and his crew to excavate under them. They were so big and heavy that they had to be broken up before they could be hauled out and disposed of. The roof of the cave was much too low to do this by swinging a sledgehammer.

Pete Smith had figured out that he could break up the rock by drilling into it and inserting blank rifle cartridges deep into the holes. He set them off by tapping a long metal punch against them with a hammer. "It's a technique that I've worked on over the last couple of years and honed down to a fine art," he told me. "It's pretty effective."

Smith donned his caver's helmet and ear protection. With help from Heaton and company, he dragged his drill and other equipment into the cave and set to work in the dank Bear Passage. Each time he was ready to set off a cartridge, everyone ducked and turned away. The sharp bang pierced the air and shards of rock went flying. But it worked. The well-controlled charges fractured the rock without disturbing the site the way any larger scale of blasting would do. The rubble had to be carried out through the area just outside the cave where the archaeologists were working, so they also had to take a break from digging. Both crews formed a staggered line, like a bucket brigade, and passed the broken pieces of rock out from the cave. The last people on the line dumped them onto a burgeoning pile of debris at the edge of the site. By the time Smith was done, it was lunch hour.

After eating, Jim Dixon and I lounged in chairs outside the yurt while he told me how he had become involved in the project here. By around 1990, he had finished the major work on *Quest* and had "concluded that the Northwest Coast and coastal migration was the most viable explanation for the peopling of the Americas." But where to look for evidence? The big problem was that so much of what had been the coast in the late glacial period was now under water. As for the areas that *had* been above water back then, most were now up in coastal rain forest where the acidic soil made for extremely poor preservation of bones and seashells. Yet these were some of the main materials that could document human presence.

It was a genuine problem. When Daryl Fedje and his colleagues probed in the forest in the Charlottes, they hardly found any bone or antler or shell. There must have been plenty at one time, since the sea level and shell-laden beaches had been that high, but the calcium-rich organic matter had all dissolved thousands of years ago. All Fedje found were stone tools and toolmaking debris. This left Dixon wondering where to look for coastal evidence.

Then Dixon met Jim Baichtal, the U.S. Forest Service geologist who proved to be so helpful to Heaton and the other cavers. Baichtal gave a talk at the University of Alaska enthusiastically describing the so-called karst topography on Prince of Wales and the adjacent islands of southeastern Alaska. Karst refers to areas of limestone bedrock with underground drainage, rock that is so riddled with holes, fissures, and cavities it looks like Swiss cheese. It is formed because the groundwater in coniferous forests is so acidic. This dissolves the limestone, often creating entire networks of "solution" caves that are characterized by the tubes through which the acidic groundwater once flowed. "Until then," said Dixon, "we had had only isolated reports of caves in Southeast Alaska. So when Jim Baichtal gave his talk, I realized, 'Ah hah, there's a possibility. A large, vast ancient system of solution tubes and interconnecting chambers.' " Inside a limestone cave, Dixon knew, there would be excellent preservation of bone, antler, and shell.

For a few years, though, Dixon got "diverted because most of the archaeology being done in Southeast Alaska was in sea caves." These were places affected by wave action, and most were at or just slightly above sea level. "We were getting reports of fantastic paintings in these sea caves," said Dixon. So they held out exciting prospects for archaeology. It took a few years for scientists to work out the sea level chronology for the region, just as Fedje and Josenhans were also doing in the early 1990s in the Queen Charlottes. After the Ice Age, in both places sea level rose rapidly and passed its present stand, stayed high for thousands of years, and then settled down again. Eventually Dixon and others realized that these caves formed at sea level did not date back to really early postglacial times. "They could not be terribly old.

"Just about then," he went on, "Tim Heaton found the first archaeological evidence here," on Prince of Wales Island. The archaeology consisted of only a few artifacts, but that was enough to get Dixon excited. Finally, he thought, they had found a solution cave that was high above the shoreline, a site with real promise. The first stone artifacts were not directly datable, but the radiocarbon age of the animal bones Heaton and Grady had turned up was intriguing. He got in touch with Tim Heaton and Forest Service archaeologist Terry Fifield to learn more about On Your Knees Cave. Then he went looking for money to support a dig.

It was not easy. Because of the cave's remoteness, it was a costly site to excavate. Travel costs were high. Everyone had to be brought in by boat. The heavy and bulky things—the generator, fuel drums, water tanks, portable cook tent, tools, and bulk food—could not be packed in and carried up the trail, so the Forest Service had to cut down a few of the huge spruce and hemlock trees to create a small opening in the forest canopy. The heavy stuff was lowered from a helicopter while the chopper hovered above the trees. This worked, but it was expensive. The budget also had to cover radiocarbon dating, which runs many hundreds of dollars per sample.

Looking back at his struggle for funding, Dixon gave a grim laugh and said it was a minor miracle that he ever raised the money to excavate the site at all. For academic archaeologists in America, he lamented, there is always intense competition for very limited money from only a few sources. And the On Your Knees Cave site had a big strike against it: the theory of early coastal occupation was controversial. He thought that Fedje and Josenhans would have had similar difficulties raising money for their work in the Queen Charlottes if they had faced direct competition from outside projects and researchers. But they were Canadian government staff scientists working within established and relatively well-funded organizations, so they had somewhat privileged access to the expensive research ships and other resources.

In Dixon's case, he first applied jointly with Heaton for a National Science Foundation grant to cover the 1996 season. These grants had to be peer reviewed. With hindsight, he could see that his first grant application was probably too ambitious. He had proposed to include some underwater survey work, which would have added greatly to the cost of the project. In the end, he was turned down on the grounds that the Prince of Wales cave dig was "entirely too speculative." Meanwhile, though, Heaton was able to land the small, two-week National Geographic Society grant that allowed him to return to the cave. "Then, in the early summer of 1996, Tim found the human remains" on the last day of his planned fieldwork.

That's when things became complicated, because the Tlingit

could have objected to having the human remains examined and dated at all. Witness the Kennewick Man kerfuffle, which unfolded in Washington State that same summer. The Kennewick skull and other bones had been found along the Columbia River. Although initially the bones were thought to represent a possible murder in modern times, the Umatilla objected to having a forensic anthropologist study them. The confrontation quickly escalated, with an Umatilla spokesman branding the anthropologist a "grave robbing motherfucker." The Indians made it clear that they wanted to take possession and reinter the remains immediately.

Fortunately, the attitude of the local Indians and the interpersonal dynamics at work on Prince of Wales Island could not have been more different. Terry Fifield, the district archaeologist with the U.S. Forest Service, flew in by helicopter the day after Tim Heaton's discovery and took possession of the human remains and the bone artifact. He promptly notified the local tribal groups, as required under the Native American Graves Protection and Repatriation Act (NAGPRA). There was a tribal division on the island. The Klawock and Craig communities on the more northerly part of the island, where the cave site was located, were predominantly Tlingit. The tribal authorities on the south end were mainly Haida. They deferred to the north island leaders.

Fifield began a series of meetings with the Tlingit to solicit their support and listen to their concerns. At first, both the Craig and Klawock tribal councils had strong reservations. As Millie Stevens, president of the Craig Community Association, later told a journalist, there was a lengthy discussion among the ten Craig board members. One concern was to avoid making the name or the location of the cave widely known. They feared that grave robbers might desecrate it, as had happened at burial sites in other parts of Alaska and elsewhere. They wanted the bones to be treated with dignity and to prevent photographs of the bones from being published.

Working in Fifield's favor, though, was the fact that the Tlingit were genuinely interested in finding out what these remains might represent. Stevens was able to convince the more reluctant board members to allow the initial radiocarbon dating of the bones. In part this was to "give us a better idea as to how long people had been in this area. We really believe that the Tlingits were the first people" here. The Craig association voted unanimously to allow the dating and excavation to proceed.

According to Rosanne Demmert, president of the Klawock Cooperative Association, both tribal groups quickly agreed to allow the skeleton to be dated, with the understanding that the bones would later be reburied at the site. The tribal councils also got a written agreement from the Forest Service that all decisions regarding the skeleton would be discussed with them every step of the way. In addition, the tribal councils were to have prior review of any new information that was to be published regarding the skeleton.

Some of this impressive cooperation can be attributed to Terry Fifield's diplomacy and personal standing in the very small rural community of Klawock, which is about 70 percent Tlingit. His children go to the same school as most of the council members' children in Klawock, Fifield recalled. "We're active on all the same committees. I see everybody in the grocery store. They know me. People trust me to some extent, and I'm personally accountable for the things I say in public. There isn't the amount of distrust that can develop in a more urban context." Fifield was readily available to attend meetings, where he could discuss the issues and smooth over any ruffled feathers. And he exchanged dozens of phone calls and letters with the tribal authorities.

There was no confrontation between science and native beliefs. Instead, the Tlingit assumed an active role. Not only did they agree to allow the precious finds to be studied and dated, they seconded young Tlingit to work on the excavation as interns. (Initially these interns were paid under a grant that Dixon obtained. Then the umbrella agency for tribal groups in the region, the Sealaska Corporation, took over their funding.)

I had discussed these developments over the phone with Fifield before coming up to visit the site. He modestly played down his own role, arguing that only the more confrontational situations between scientists and Indians ever make the news. Where there is good co-operation, he said, you never hear about it. This was what I had also witnessed in the Queen Charlottes, where Daryl Fedje worked so well alongside the Haida. There, too, the Haida had been granted major roles in the archaeology projects and some real veto power. The personal trust and spirit of joint endeavor on Prince of Wales Island were entirely different from the poisoned atmosphere of the Kennewick Man controversy. In a book primarily about Kennewick Man, David Hurst Thomas, curator of the American Museum of Natural History, extolled the Alaska situation as a "model" for how archaeologists and Indians should interact.

With the Tlingit on side, Jim Dixon was able to plunge in. "Terry Fifield phoned me in Denver, and I flew up here right away." They hiked up to the cave, crawled in and "checked it out." The tribal councils had already given permission to date the human bones, the mandible and pelvis, "which I took back with me to Denver."

For that crucial task, Dixon turned to Tom Stafford, one of the top radiocarbon dating experts in America, who has a private lab in Boulder, Colorado. (Stafford was involved at the same time in dating human remains from California that would also play a key role in the coastal migration story. But that's getting ahead of the tale.) "Using a micro-drill, Stafford and I took samples of bone from the mandible and hip." Obtaining and dating samples from both was important. "We wanted to see if it was the same individual, which would be likely if the dates came out to be very close. Also, the more dates, the better the validity." Their expectations were confirmed. "Based on the similar degree of staining, size, and estimates of age and gender, all the human bones appear to be from the same individual."

Initially, the dating showed the mandible and pelvis to be about 9800 years old, "making them the oldest human remains ever found in Alaska or Canada. That was very exciting. It clearly warranted further work. And we got permission to continue. The tribal governments endorsed it." Eventually, the 9800 year date had to be modified, but this was for reasons that added significantly to the knowledge derived from the human bones. Meanwhile, the bone artifact that had first caught Tim Heaton's attention turned out to be about 10,300 years old. That put it back into the end of the Ice Age and made it the oldest artifact found thus far on the North Pacific coast. Even before further excavation began, On Your Knees Cave was a landmark site.

The next fall and winter, Dixon drafted a new proposal for a major National Science Foundation grant, requesting $300,000 for three years. This time he left out the marine survey but included a request for money to fund two native internships. These were to be divided among four young people from Klawock and Craig. It was a relatively large funding request, mainly due to the remoteness and logistics involved at the site. His grant was approved, but at first only to the tune of $100,000 and for only one year. The very old artifact and the fact that ancient human remains had been found apparently tipped the scales in his favor. Dixon's peers in the archaeological mainstream still had reservations about coastal migration, however. "They said, 'we'll see what you find,' " and Dixon could understand their hesitancy. "At that point, there had only been a few artifacts, and we didn't know if there were more." But the first summer of the dig yielded enough exciting evidence for him to go back and request $200,000 to complete the project. Even then, though, getting this second batch of money was a close call.

As Dixon explained it, projects may be approved, but that doesn't necessarily mean they will be funded. Following approval, they are graded with "A's" and "B's" to determine which ones will actually receive money, and how much. A project just about needs straight A's to be high enough on the list. Dixon's proposal got two A pluses, one A, and one B. "And a B is almost the kiss of death," he said. So at one point it looked as though he would fail to receive the money to continue. In the field of archaeology, "you're competing with sexy projects like the million-year-old hominid research in Africa. So, it depends on the year you apply. If there are five or six major hot projects that pop up," he laughed ruefully, "you're dead. It's very competitive."

Fortunately, though, in the course of the application process he found out that there was a separate little pot of money earmarked by the National Science Foundation for "polar programs." Even though the region was not actually polar, it qualified due to the Alaska location and the tie-in with glacial history. The needed $200,000 was secured.

Work at the cave site began in 1997. The staffing was mainly by young volunteers from the Lower Forty-Eight states, most of them students from universities in Colorado, plus the two positions allotted to local native interns. In later years, the Sealaska Corporation, which is owned and run by the tribes of the region, stepped in and funded the two native intern positions itself. By relying largely on unpaid volunteers, and with the help from Sealaska, Dixon was able to stretch the money and extend the dig into the summer of 2000.

As I could see already, it was turning out to be a highly productive dig. From the start, Dixon and Heaton knew they were onto a key site. There was now solid evidence that people had been on a remote offshore island at least a thousand years earlier than previously thought, and earlier than anywhere on the adjacent mainland.

The coastal migration scenario was looking better and better.

CHAPTER TWELVE

Boulders That Talk

AS THE RESULTS from the Alaska cave trickled in, paleoanthropologists began to acknowledge that people had been living on islands off the North Pacific coast considerably earlier than they had imagined. Meanwhile, major additional support for the coastal migration scenario came from related developments in two areas far from the Northwest coast: in the region of the much-touted ice-free corridor and in distant Chile.

Knut Fladmark had described the corridor as a myth "running through the minds of many archaeologists, like a highway beckoning Paleoindians south from Beringia." Initially, in the 1930s, the corridor was envisaged as a long, narrow region that opened up to animals and people only very late in the last glacial period, as the Laurentide and Cordilleran ice sheets melted and drew back from each other. Glaciologists generally supposed that at the peak of glaciation the two great northern ice sheets had coalesced, leaving no viable migration route. The path people might have taken at the tail end of the glaciation ran eastward across Alaska up the valley of the Yukon River into the watershed of the Mackenzie River, and then southward along the eastern slope of the Rockies.

In the 1950s and 1960s, as the Clovis First concept became entrenched, some archaeologists began to argue that the corridor must have existed throughout most, even possibly all, of the last glaciation. This was based on precious little hard data. Whether the ice sheets had in fact coalesced, effectively blocking migration (and if they had, precisely when and for how long) remained unanswered right through the 1980s. Part of the difficulty was that the corridor ran some 2000 miles through a variety of geographical zones. Glaciers could be advancing in one area while retreating in another. Yet for the corridor to be a practical migration route, all of it had to be open at the same time.

Gradually, geologists studying cores and moraines in the region assembled strong evidence that the putative corridor was blocked, at least along some sectors, for nearly all of the glaciation. Alejandra Duk-Rodkin of the Geological Survey of Canada focused on the watershed of the Mackenzie River in the Yukon and northern British Columbia, tracing the complex advances and retreats of the ice for the northern portion of the corridor. In some areas, when the Cordilleran glaciers retreated toward the west the Laurentide ice sheet advanced from the east at the same time, thereby keeping the potential corridor blocked. Overall, she concluded that no gap existed between the ice sheets until about 12,000 years ago. And because ice-scoured, newly deglaciated areas are so barren of plants and animals, no viable, life-sustaining migration route existed for perhaps another thousand years.

As Knut Fladmark had pointed out, even where there was a gap between ice sheets, because of the cold winds, scoured ground, and meltwater lakes, it might not have been a place where plants, animals, and humans could have lived for extended periods. Ancient people could not have slipped through quickly with backpacks of freeze-dried food. To support a human migration there had to be well-established food resources, such as grazing animals to hunt and edible herbs, roots, and berries to gather. Carole Mandryk, a Harvard anthropologist, studied pollen and other biological indicators from deglaciated areas and constructed a paleoecological model of conditions in a long, narrow region of that kind. She calculated the potential food resources that people might have relied upon there: the amount of directly edible plant biomass for people and the stock of prey animals that could have lived off other available plants, to be consumed in turn by people. Her verdict was that there would not have been enough available nutrition to sustain a human population between 18,000 and 13,000 years ago. After around 12,000 years ago, food biomass would have become increasingly abundant. Mandryk's studies led her to conclude that the ice-free corridor "was not a possible route prior to around 12,000 years ago." Eventually, she became a vocal supporter of coastal migration at professional conferences and collaborated on research with Daryl Fedje, Heiner Josenhans, and Rolf Mathewes.

The most decisive debunking of the corridor, though, came from study of the route's southern portion.

Strung out along the eastern foothills of the Rocky Mountains is a long line, or "train," of unusual boulders called the "foothills erratics." They extend about 360 miles in all, from the McLeod River in Alberta down into northern Montana. By definition, like all erratics, they are not of locally formed rock. They were brought to their present locations by flowing glacial ice, and then left where they now lie when the glaciers melted. These particular erratics, numbering in the thousands, are angular blocks of extremely hard quartzite. Some are colossal. They range in size from a few feet long up to 135 feet for what is the world's largest erratic, the so-called Big Rock, or "Okotoks" in the Blackfoot language, which weighs about 16,500 tons. And they are the most prominent features in the largely treeless, open, and rolling countryside of that region. Because they are so conspicuous, these erratics have long been objects of curiosity. But it was the distinctive way they are distributed that made them of particular interest to geologists. Instead of being spread widely or randomly across the landscape, like other erratics left behind by the melting of the great northern ice sheets, these boulders all lie in a very narrow band, hardly more than a mile at its widest.

One scientist whose curiosity was especially piqued by them was Lionel Jackson of the Geological Survey of Canada. "The native people of the area had legends about them," Jackson told me over the phone. "They were nomadic people who, naturally, took notice of the landscape." In the case of the foothills erratics, he said, you can stand on a hilltop and see the boulders running off into the distance in a narrow line. The Indians sought to account for this. "One legend has it that the Creator, who was called Napi in the Blackfoot language, had a buffalo cloak." One day, he put the cloak down on a single enormous rock. "Well," said Jackson, warming to his tale, a buffalo cloak was a very valuable object. "The rock came alive and ran off carrying the cloak. And as it ran along, pieces fell off the giant rock, leaving the erratics train in its wake." In another story, Napi set down his cloak on the Okotoks erratic, asking the Big Rock to hold it for him. When it began to rain, Napi wanted his cloak back. The rock refused, so Napi grabbed the cloak and walked away, but the angry rock rolled after him, forcing him to run for his life. Napi's friends, the deer, bison, and pronghorn, tried to help by running in front of the rock, but it rolled right over them. Napi's last chance was to call on the bats for help. They dove on the rock and hit it just right, stopping it, saving Napi's life, and creating a crack that can still be seen today.

While out geologizing among the erratics, Jackson would often climb up on top of a large one to enjoy the panoramic view while he ate his lunch. And he would think to himself, *if only you could talk, what stories you could tell.* "Well," he said, "we found a way to make them talk." And what he learned from the rocks pretty well settled the debate about the ice-free corridor.

In the 1960s, Jackson's predecessor, GSC geologist Archie Stalker, traced the erratics north and west to the source of the quartzite. The trail backward led far up the Athabaska River to a mountainside high in Jasper National Park. The Athabaska at that point flows northeastward out of the Canadian Rockies. During glacial periods, valleys that now contain rivers like the Athabaska were filled with flowing "montane" glaciers. The erratics had either been torn from the mountainside by the ice, or else they had fallen down onto the ice surface during landslides. Then the glacier transported them on or near the surface of the ice, as on a conveyor belt, down the Athabaska and into the lower foothills. At that point, geologists reasoned, the glacier carrying the erratics must have run smack into another great ice sheet, the Laurentide, coming from the east. Where they met, the geometry and pressure from behind squeezed the ice at the forefront and made it change its direction of flow southward, a bit like pinching a wet cherry pit between your fingers. Thus, the glacier flowing northeastward down the Athabaska made a sharp right turn and flowed southeastward, parallel to the mountains, carrying the erratics with it. In other words, at the time when the erratics were deposited there must have been a coalescence of ice sheets along the eastern side of the Rockies. Otherwise—if there had not been a wall of ice coming from the other direction—the glacier descending the Athabaska Valley would have fanned out to the east. When it melted, the erratics would have been deposited in a much broader and spreading pattern west of today's city of Edmonton, Alberta.

Then came the crucial part of the geological analysis. The erratics were deposited during a glacial advance in which the ice sheets coalesced. But there were several major glaciations during the Pleistocene, and the idea was to figure out which one had left behind the erratics. The geologists realized that no subsequent advance of glacial ice from either direction could have gone beyond the line reached by the ice fronts at the time the erratics were deposited. If they had, then one or another of these more recent advances would have bulldozed against and over the erratics haphazardly and disrupted their neat line of distribution. A more recent glaciation could have been somewhat *less* extensive, however. In such a scenario, the erratics might have been left by one glaciation, and a later glacial advance (let's say the most recent one) would have left the erratics where they are found today. If this latter scenario were correct, there would have been a "corridor" between the ice sheets of that most recent glaciation.

If geologists could figure out roughly *when* the erratics were transported, they might be able to settle this issue in the debate on the peopling of the Americas. Fortunately, a brand new technology became available, and Lionel Jackson launched a research project to exploit it.

In late October 1995, two of Jackson's colleagues, Kazuharu Shimamura of the Geological Survey of Canada and Edward Little of the University of Alberta, drove out into the Alberta foothills to take samples of the erratics. The quartzite had natural "bedding planes" in it, faults that in some cases allowed them to pry loose pieces to put into their sample bags. For some of the samples, though, they had to use a gasoline-powered portable diamond drill, which one of the scientists held while the other pumped water to lubricate the bit. The weather had turned so cold that the drill bit kept freezing up on them, and they had to fetch antifreeze to add to the water. In the end, though, they collected half-kilogram samples from eight of the erratics. Jackson had these sent off to two specialized labs in the United States for so-called "cosmogenic chlorine 36 dating."

The principle behind this technique is that atoms of this particular chlorine isotope accumulate in rock due to bombardment of the surface of the rock by cosmic rays. The amount of this chlorine isotope can be taken as an approximate measure of the length of time the rocks have been exposed to the open sky. Jackson assumed that in most cases the foothills erratics were initially buried in a mountainside, at least until a landslide deposited them on a moving glacier. He further guessed that, when the ice slowly melted, most of the boulders ended up in stable positions and did not subsequently roll over. The gently sloping ground in the foothills lent itself to such stability. If both assumptions were true, then chlorine in samples taken from the top surface of the rock should reflect the length of time since the erratics were transported from the mountains.

It's a highly sophisticated method, Jackson told me. "We're talking about a very, very minute amount of material, roughly one atom out of a number with sixteen zeros after it. If you take a sample of say 500 grams, it might have 6000 atoms in it of that particular isotope of chlorine." So the first step was to separate out this isotope, which was done by the first lab. Then this minute bit of chlorine 36 was sent to the second lab and put into a particle accelerator, where the atoms were counted. Finally, the results went back to the first lab for analysis by a special computer program.

Seven out of the first eight erratics samples fell into the 12,000 to 20,000 year range. A year or two later, additional samples were collected from eleven more erratics, and tests on these confirmed the initial results. All of them fell into the 12,000 to 18,000 year range. This meant that the foothills erratics must have been deposited by the most recent glaciation.

Since the ice sheets had to have coalesced during that time period, at least in that southern region, the implication was clear. "These ages eliminate the possibility of an ice-free corridor during the last glacial maximum," Jackson and his coauthors wrote in 1997. "Because the foothills erratics train is a byproduct of the coalescence between montane and continental ice, a late Wisconsinan age for it argues strongly against the entry of humans into the Americas from Beringia during the climax of the last glaciation via an ice-free corridor, a long-standing hypothesis. It is also consistent with the growing evidence that the last continental (Laurentide) ice sheet was the most extensive one ever to exist between the international border and the Arctic Ocean in western Canada." Jackson concluded that any corridor could not have opened until around 11,000 to 11,500 years ago.

The findings had implications for the coastal migration theory. A corridor that opened 11,000 to 11,500 years ago might account for people making it south of the ice in time to leave the Clovis artifacts, which date to that period. But what if people were south of the ice significantly earlier? With the corridor obstructed until around 11,500 years ago, there would have to have been another way to get past the ice. Jackson's research would then be the final nail in the coffin of the corridor as the route taken in the first peopling of the Americas.

It was a big "If." Had pre-Clovis people really reached beyond Beringia? The answer was provided by a field conference as significant as the one held at Folsom in the late 1920s. And it took place many thousands of miles to the south at Tom Dillehay's long discussed and debated Monte Verde site in Chile.

The boggy site first came to light in 1976, when fractured bones, a biface stone tool and some modified pieces of wood were found to be eroding out of a stream bank. The following year, Dillehay assembled a team and began an excavation there that extended intermittently over nearly a decade. What he and his coworkers unearthed included unusually well-preserved organic materials that rarely survive intact down through the millennia. The finds included wood, reeds, seaweeds, animal skins, butchered mastodon bones, and meat. There were even fossil "coprolites," shriveled and dried-out human feces that look, well, just as mummified turds might be expected to look, and that can contain seeds, pollen, and eggs from intestinal parasites, all of which help archaeologists to reconstruct the diets of ancient people. The organics were preserved in this case because soon after the site was abandoned, the water table rose covering the spot in waterlogged peat, which cut off the oxygen that would allow bacterial decay. In addition, there were objects and materials that more commonly survive, including more than 700 stone artifacts and a number of fire pits. Most poignant of all was a child's footprint in the mud. All were excavated in an undisturbed geological context that lay buried under strata of peat that could themselves be dated. The large number of organic items permitted many radiocarbon dates. Some were in the 13,000-year range, but on average they clustered around 12,500 years of age.

Dillehay reconstructed an encampment that included twelve wooden frame structures with remnants of their hide coverings. The site was apparently "home" to between twenty and thirty people. Among the artifacts were smooth, rounded stones that were probably used to kill small animals. The seaweed bespoke visits to the coast, a day's march away, possibly to harvest shellfish as well. There were also the remains of forty-five varieties of edible plants, some of which do not grow less than 150 miles from the site today, which seemed to indicate fairly widespread travel or trade networks. Taken together, the picture was of people living a diverse life of hunting and gathering.

The mastodon remains attested to the existence of large game, and the cool, wet climate was not very different from what the Clovis people might have experienced in some parts of North America more than a thousand years later. Still, on the whole, Monte Verde did not fit the big-game hunting image that had come to be so closely associated with the late Ice Age. That's one reason why, when Dillehay began to release his preliminary results, he was met with great skepticism.

It was not just the Clovis Mafia who had their quibbles. Richard Morlan of the Canadian Museum of Civilization (one of the butchers of Ginsberg the elephant) was himself a believer in a relatively long chronology, at least for Beringia. In a 1989 review of Dillehay's first book describing Monte Verde for the journal *Science*, Morlan wrote: "The majority of the stone artifacts (90 percent) were said to be pebbles picked up from a creek bed and minimally modified by use." In other words, they might not be artifacts. "The creek bed context itself raised the specter of fluvial deposition for the remarkably well-preserved wood and the scattered mastodon bones." In other words, they might have floated in from elsewhere. "Hut foundations made of small logs staked to the ground were reported, but this skeptic wondered whether the logs were naturally fallen trees with the 'stakes' actually representing branches preserved on the underside of the trunks."

On balance, though, Morlan, like Jim Dixon and a growing number of American archaeologists, seemed convinced that Monte Verde was indeed a valid site that represented human occupation in southern Chile 12,500 to 13,000 years ago. "My skepticism was diminished and my interest aroused when . . . I heard a conference paper in which Dillehay explained that the stakes were made of a different species of wood from the logs through which they were driven and that some of the stakes were wrapped with reeds." But for almost another decade, the orthodox mainstream continued to doubt Dillehay's results, suspecting that somehow either artifacts from more recent times had been washed down into deeper and older layers, or that his dates were wrong because carbon from nearby volcanoes might have contaminated the site.

Typical were the arguments mustered as late as 1995 by R. G. Matson and Gary Coupland in an overview of northwest coast prehistory: "Monte Verde . . . is inherently unlikely, in suggesting an 'Archaic' life way [more generalized hunting, gathering, and foraging, as opposed to a focus on big game] earlier than any in the Old World. The artifacts associated with this site are very dubious. Of course, being in Chile, it is not immediately accessible to other North American field archaeologists. We expect that Monte Verde will eventually be rejected as more information becomes available."

But Dillehay enjoyed continuing support from the Smithsonian Institution, which was preparing to publish his long-awaited 1300-page final report on the dig. Scientists were either going to have to accept his findings, or explain why not. So a blue ribbon panel of nine archaeologists, led by Alex Barker, curator of the Dallas Museum of Natural History, joined him for a traveling field conference in February 1997. Accompanied by a small media contingent, they included some of the most prestigious and active scholars of the early peopling, such as Dennis Stanford of the Smithsonian, David Meltzer of Southern Methodist University, and James Adovasio of Mercyhurst College. And there were two leading members of the "Clovis Mafia," Dena Dincauze of the University of Massachusetts at Amherst and C. Vance Haynes of the University of Arizona.

First they viewed artifacts from Monte Verde at Dillehay's home university in Kentucky. Then they flew to Chile and visited the Universidad Austral to see more artifacts and review the related geological and paleoenvironmental context. Finally, they poked around for a full day at what remained of the site itself, which had been disturbed by local development. They were able to see for themselves the kinds of strata from which Dillehay had unearthed his artifacts and environmental indicators. They could search for possible alternative explanations of how such clearly dated artifacts might have become inserted into the sediments below the well dated (and later) layers of peat. In the end, no satisfactory alternative to Dillehay's interpretation could be found.

"I was the heavy," recalled Haynes, probably America's most respected geochronologist and Clovis First supporter. But he saw the artifact-bearing layer of soil lying in a "secure stratigraphic context" under a layer of peat that dated to 10,300 to 12,000 years old. "So, the artifacts had to be older than that, and I had to buy those dates."

Others were more impressed by the unquestionable nature of certain artifacts. Among the most remarkable were poles and tent stakes that were lashed together with knotted cords made from local grasses. "That's something nature doesn't do," said Alex Barker, "tie overhand knots."

And so they retreated to the La Caverna saloon in the coastal town of Pelluco to question, challenge, and debate their conclusions. Meltzer polled the others on whether they could agree that Monte Verde was 12,500 years old. The group voted on it, and it was unanimous. Barker, the delegation's organizer, raised his bottle of beer. "I'd like to propose a toast," he smiled, "to the passing of a paradigm."

The acceptance of Monte Verde made headlines around the world. "It totally changes how we think of the prehistory of America," Stanford told the *Washington Post.* "Our models clearly are not right." Dincauze called it "a new benchmark in knowledge." And none could ignore what the date of this site in far-off South America implied, especially if the corridor east of the Rockies was indeed blocked until a thousand years later. There must have been another way that people got past the ice. As Harvard's Carole Mandryk told the *New York Times* science writer who covered the Monte Verde developments, "They came down the coast. I don't understand why people see the coast as an odd way. The early people didn't have to be interior big-game hunters, they could have been maritime adapted people."

CHAPTER THIRTEEN

Needle in a Haystack

DARYL FEDJE WATCHED from the *Vector*'s aft deck as the ship's crew sent the big clamshell-like jaws down into the water. From what he knew about how sea level had changed late in the Ice Age, any artifacts dredged from the bottom here would probably be over 10,000 years old. It was 1998, and this was the first really promising opportunity for Fedje and Heiner Josenhans to hunt for undersea artifacts in the Queen Charlotte Islands. They had been allotted four full days of ship time on the *Vector*. Unlike the almost perfunctory attempts at dredging I had witnessed earlier, they now could devote almost all their time to attempting to recover the stone tools that might prove people had been here in the last few thousand years of the glaciation. And unlike the previous cruise, they now had the ability to target likely ancient habitation sites on the sea bottom with great accuracy.

That greatly enhanced capacity was thanks to new technologies that had only become available since the earlier cruise. It was a development that I had followed ever since a surprise visit by Josenhans.

"Wait till you see this video," he told me. "It will just blow you away." Josenhans had dropped in on me and my wife Annie at home in B.C. It was the summer of 1995, a year after we had met during my first trip to the Charlottes with the *Vector*. Back at his home base in Nova Scotia, he was involved in a project surveying and imaging the seabed for a proposed undersea fiber-optic cable. That research was applying new digital technology. The video was a sort of animated motion picture constructed from a long sequence of images, each one incrementally different from the next. In techno-speak, it was based on a series of "digital elevation models," which were really just high-resolution contour maps of the seafloor. It was technology like this that he'd just love to apply to the undersea search in the Charlottes, he said. He lent us the video to watch at our leisure and mail back to him afterward.

When we popped the tape into the VCR, we saw why Josenhans was so excited. The effect was spectacular, just like flying underwater. The screen displayed a view of the seafloor with colored shading and contour lines that made it almost three-dimensional. As the video opened, the imaginary camera was viewing the scene from a stationary position just below sea level in shallow water. Stretching out ahead was the cable. It ran along a twisting path between various humps and bumps on the bottom and followed the seabed down into deeper water. This increasing depth was indicated by ever darker shades, first of blue, then purple.

The virtual camera started to move, following the cable down along its tortuous path, which represented the crossing of a deep strait. Down we swooped, into sinuous canyons that snaked along the undulating terrain and provided a convenient cable route, while drowned cliffs rose along the sides. At one point, an ancient river system, carved out when sea level was low, cut its way through the cliffs and fed into the canyon. Meandering tributaries fed into the main riverbed from each side, like branches attached to the trunk of a tree. At other points, there were little hills or ridges, apparently representing moraines or drumlins that had been left by glaciers. Up over slight rises we went, and then down again farther into the deep. As we approached the opposite side of the strait, the virtual camera followed the cable up, up into brighter water until we reached the surface. Whereupon it turned and took us back down again, faster this time, zipping across the strait to our starting point, with all the undersea features clearly and vividly delineated.

I had watched Josenhans struggle as he pieced together black-and-white paper printouts from the sidescan sonar on the *Vector*, trying to reconstruct the bottom contours and locate the river channels and ancient lakes. These colored digital images made it just so much easier to grasp the relationships and scale of features on the sea bottom. No wonder he and Fedje thought digital imaging was the magic bullet that would vastly improve their chances of finding ancient undersea occupation sites in the Charlottes. It could provide much clearer maps of where the beaches and stream deltas had been when sea level was low. Then they could try to find artifacts by precisely targeted dredging.

Besides, what were the alternatives? All the other ways of looking for artifacts underwater were pretty well ruled out by cost, safety, or logistics. Robot submersibles were a possibility. But as I had seen, the small ones like Henk Don's MURV were not powerful enough to contend with swift currents. A sizable mother ship had to be positioned above at all times, both to deploy them and to provide power and communications links. On balance, they were expensive to operate, and their utility for archaeology was doubtful. If a submersible like MURV tried digging with its robot arm, it would only stir up the sediment and blind itself. Archaeologists could not really expect anything interesting and recognizable to be lying exposed on the sea bottom in any case. Don himself had admitted that he could not likely use MURV to spot artifacts.

Finally, there was undersea excavation by scuba divers. That had been tried by Norm Easton's group at Montague Harbor in the early 1990s, and it had worked. But even in shallow water and very close to shore, the logistics of running such a dig were not simple. There had to be an airlift or water dredge system powered by a generator or compressor on shore. If they operated farther from shore, it would have to be floating above the dig site on a barge, which would add an extra layer of complexity and cost. Even in shallow water, divers could not realistically work more than forty minutes to an hour at a time before becoming tired. So two or more shifts of divers, plus support personnel, were required.

Nevertheless, Fedje decided to follow up on the reconnaissance dives made by Kim Conway and Pete Waddell during the 1994 *Vector* cruise. In the summer of 1995 he selected a couple of sites in the Charlottes and had a team of divers try an underwater excavation. At the time, one of Fedje's large survey crews was operating in the area with a fully equipped central shore camp, so it was practical to send the divers out each day by Zodiac.

The project involved two professional Parks Canada divers, Jim Ringer and Charles Moore. Moore had taken part in Easton's project. They were teamed up with two certified Haida divers. One was Bert Wilson, who also headed up the Haida team of trainees working with Fedje on the summer surveys. The other was the irrepressible Tom Greene Jr., the tree faller who had cleared the trees from Sgan Gwaii village almost fifteen years earlier. They had allotted about two weeks for the work, but time just seemed to slip by before they even got started. Bad weather, delays getting the equipment into place, weekends off, and lost time due to illness all slowed things down. The first test pit they tackled was in about fifteen feet of water at Echo Harbor on Moresby Island. It took several days just to get properly set up on shore and stake down the essential datum or reference line along the bottom. Everything was to be measured from this. At last they donned their dry suits and tanks and began to dig in the middle of a small, clear patch of silt and small angular stones.

As Tom Greene described it, each diver would dig by holding a trowel in one hand and a suction hose in the other. This so-called water dredge, which was powered by a water pump on shore, sucked up all the loose and small stuff and shot it into a holding basket set into a big garbage can on the surface. Daryl Fedje was there on shore most of the time to do the screening.

The first thing the divers discovered was that the trowel work stirred up the loose sediment so much they could hardly see. "The visibility was zero most of the time. It was mostly by feel," said Greene, when he told me about the experience a month or two later during the totem pole project. As Charles Moore noted in his telegraph-style official report, "very bad visibility hampered set up, improved slightly for second dive of day, much better for third."

Still, the system worked. When they got deep enough that soft material began slumping in from the sides, they set up a small, square caisson made of plastic in the middle of the hole and kept digging. Eventually they dug their first hole down about four feet. And right in that hole Jim Ringer found a stone object that Fedje decided was probably a club handle that had been shaped by grinding. It reminded Greene of the handle of a very large Buck-brand woodsman's knife, with the same spacing of humps and hollows for finger grips. Judging by the style of the club, it was fairly modern and had probably been dropped from a canoe just off the beach within the last few hundred years. The baskets of sucked up material were brought ashore to Fedje for screening. By the end of the first day, he had also found some stone flakes and a possible microblade core, which was highly promising.

After a couple of days, they moved the dig farther out along the datum line and tried again in slightly deeper water, where anything they brought up might be a bit older. Trying a new spot also might improve their chances of scoring a hoped-for lucky hit, such as finding a rich cache of artifacts or perhaps a fire hearth or fish trap. But this time they were almost skunked; only one definite flake stone artifact was found in the second hole. During each shift, one Parks Canada diver would work with one of the Haida. After about an hour, the other pair would take over. But the work continued to go slowly, and did not bring up anything of great importance. As Moore's notes summarized, "The stone handle is likely no more than 200 years old and likely represents loss or discard. The other stone artifacts are likely 'trickle down' from the intertidal old site on the nearby beach." To Fedje they looked like what he would expect to find in the typical 9000-year-old range. Without a lot of further work and analysis, he could not exclude the possibility that they had simply been washed down into the water from today's shoreline.

Then Fedje moved the operation to another area that he thought might be productive, near Klunkwoi Bay at the mouth of Anna Inlet in water that was slightly deeper than at Echo Harbor, which again meant older. "It's a really interesting spot," he told me. "An old river delta at the lower sea level. There's a nice terrace there that's about thirty to forty feet below sea level today. If they found something, it would probably be a couple of hundred years older than what's on the beach." That might push his record of known human occupation in the Charlottes back to 9500 or 9600 years ago, which would be quite a significant accomplishment, so hopes were high. The divers worked two half days digging in a single test hole, but they found no definite artifacts.

Given all the careful planning and effort, it was with considerable frustration that they gave up when the allotted time ran out. With government work schedules and the logistics required to support such a dig, there was no flexibility to carry on. In all, they had put in about six reasonably effective diving days, and the results were slim. As with the robot submersible, this kind of undersea effort has not been tried again in the Charlottes. For Fedje and Josenhans, dredging with the grab jaws was the only way to get at really old stuff deep underwater. The first attempts had been crudely targeted. But their prospects were improving.

By the mid-1990s they had made great progress on the fundamental marine geology and mapping. The 1994 *Vector* cruise pretty well filled in the missing points on the sea level curve right back into the last glaciation. They knew quite precisely, that is, where sea level stood at any given time on the eastern side of the Charlottes and out in open Hecate Strait. They had also identified all the undersea areas that would have been late Ice Age steppingstones: Laskeek Bank, Dogfish Bank, Middle Bank, Goose Bank and Cook Bank.

Next, they obtained time on a larger research ship, the *John P. Tully*, which was equipped with the vibracore system that Josenhans had so sorely missed on the *Vector*. Those cores enabled them to pinpoint something they had not been certain existed: a sheltered body of water on the eastern side of the Charlottes where the deepest bottom areas had been dry, exposed land right back to around 12,500 years. The area was Juan Perez Sound. Until that 1995 *Tully* cruise, they thought it might be impossible to follow sea level downward (and therefore backward in time) beyond about 11,000 years without getting out into exposed Hecate Strait. Trying to do marine archaeology in such wild waters, they knew, would be extremely difficult, if not impossible. After that cruise, they had a highly promising place to search for undersea artifacts and environmental indicators that extended significantly farther back into the last glaciation.

The only thing missing was much better mapping and imaging technology. The previous technology, sidescan sonar, generated a black-and-white printout showing shadowy features on the sea bottom to each side of a line that represented the ship's track. The exact depth of each feature was not indicated. It was difficult for geologists to piece together an overall image of the terrain from the long, narrow printouts. And it was impossible to correlate the images precisely to the location of the ship.

All this was changed with the introduction of swath bathymetry (also called multibeam sonar), a sophisticated system in which a special surveying vessel sends out numerous sonar beams as it cruises through the water. Each beam is angled downward and outward in a fan shape radiating sound to either side of the boat's keel. Each individual returning echo is distinct from the others, and the depth of the bottom at the point hit by each of these beams is calculated by a computer. The system acquired by the Canadian Hydrographic Service in the mid-1990s sends out 127 beams at a time. The end result, once the computer has analyzed the data and combined it, is an extremely accurate and easy to comprehend image of the sea bottom, much like a topographical map complete with contour lines. In water of 300-foot depth, for example, it covers a swath about 900 feet wide. The image at that depth is like a photograph made up of pixels, with a resolution of roughly one pixel for every four square yards of sea bottom. Depth is accurate to around one foot. And the picture is in false colors that shade into each other and give a three-dimensional sense of perspective. (Like the video Josenhans had lent to me, light yellow shallow areas shade gradually into greens at a middle depth, then into blue and finally into a deep purple at even greater depth.)

The computerized image is linked precisely to the ship's location, as determined by the GPS satellite system. In an important simultaneous development, during the mid-1990s the GPS itself became much more accurate. The system of navigational satellites had been put into orbit by the United States to serve its military interests. In principle, it was accurate to within a few feet, but initially only the U.S. armed forces could take full advantage of that. An error factor had been intentionally introduced to the system, so that all non-U.S. military users would have considerably less precision. The idea had been to deny adversaries, such as Soviet missile submarines, the full advantages of such deadly accuracy. This intentional inaccuracy had been responsible for some of the difficulty the *Vector*'s captain had locating targets for sediment cores in 1994. By the later 1990s, though, the United States had greatly reduced or eliminated this error margin. Now a ship equipped with swath bathymetry and GPS could drop a core or grab with a precision of ten to fifteen feet.

Josenhans successfully lobbied the Hydrographic Service to send its new and wonderful little vessel to do a quickie survey in areas he and Fedje identified. The most important ones were in Juan Perez Sound. Once the survey had been done, and the data stored in the computer, any ship could return to the area and, guided by its own GPS, apply that survey data.

This allowed Fedje, McSporran, and Josenhans to return in May 1998, again on the *Vector*, and make their first concerted attempt to dredge up artifacts from the bottom. Unlike the earlier cruise, when the main goal was marine geology and dredging was tried only in spare moments, this time the focus was on using the grab to sample as much of the sea floor as possible. Instead of having a single person doing the screening, they had three screens hanging over the side, each with a powerful hose to wash down the dredged material and an archaeologist to pore over what was left on the screen. On board to assist Fedje and McSporran with screening were Parks Canada staff archaeologist Ian Sumpter and the Haida trainee whose sharp eyes and experience Fedje most admired, Jordan Yeltatzie.

Based on the swath bathymetry images, they focused on a corner of Juan Perez Sound called Werner Bay. This was just outside Matheson Inlet, an area where Fedje had earlier tried a little dredging, but without success because the surface sediments were too thick. Along a creek feeding into Matheson Inlet, Fedje had found one of his key sea level artifacts, the stone tool with barnacle attached that proved to be 9200 years old. That same drainage system, which followed a depression carved by a long-vanished glacier, snaked its way out through Matheson Inlet and along the bottom of Werner Bay. Using the new imaging technology, Fedje and Josenhans could clearly see level terrace areas on the sides of that ancient and now-drowned stream. A series of them followed the drainage system as it meandered along the bottom at varying depths. Each one seemed a likely ideal camping spot for people at different times in the distant past, depending on where sea level had stood. These were their prime targets for dredging.

The captain was able to position the ship accurately over each target. The effort was efficient and relentless. Down went the big jaws. Up came the scoops of gravel and shell, of rock and sand and mud. The deck crew shoveled each jawful of bottom material into buckets and big plastic boxes, labeled it, and carried it over to the screeners. Hour after hour the scientists and deck crew worked. One day of dredging and screening came and went. Then a second. The screeners blasted each sample of bottom muck clean and picked through it carefully, looking for that elusive trace of human presence. But nothing appeared.

By late afternoon on the third day, with only a fourth day scheduled and no other ship time at their disposal for that year and no success so far, Fedje and Josenhans were becoming a bit despondent. Fedje went off to fix the refrigerator at his onshore campsite. (The ship was being shared with fisheries scientists working in the area, so there was not room for everyone to sleep on board.) This left McSporran, Yeltatzie, and Sumpter at the screens. Josenhans was up on the bridge. The ship was hovering 175 feet above a place that would have been a perfect habitation site 10,200 years ago. It was a flat spot of land just behind a beach. A sand spit hooked around to the south and east, sheltering the campsite from the prevailing southeasterly wind and waves. A small river would have flowed out of the dense forest just to the west, fanning out into a delta. This should have provided any people living there an ample supply of fresh water and a place to build fish traps. There would have been shellfish beds close at hand and lots of edible seaweed as well.

McSporran was washing down the latest screenful of bottom dredgings and sifting through the little pieces of rock, shell, and bone. An enthusiastic Coast Guard deckhand was helping her, shoveling the muck from the plastic boxes onto her screen. Suddenly it happened. "Gee, what's this?" he asked, picking up a distinctive piece of stone and handing it to McSporran. It was a piece of glassy black rock about four inches long, with razor sharp edges. She turned it over and stared at it, hardly believing her eyes. Could this be it, she wondered, the needle in a haystack? The first underwater evidence of ancient coastal dwellers living here as far back as the last centuries of the Ice Age?

Word of the find filtered quickly through the ship. Sumpter and Yeltatzie stopped screening and hurried over to look. Josenhans heard the news up on the bridge and said to himself skeptically, *Yeah, right!* They had dredged up so many samples, and there had been false hopes before. So he told himself, *No, I'm not even going to pay any attention. It's too good to be true. I'm not going to get suckered.* Eventually, though, curiosity won out and he went down to join the growing group of scientists, ship's officers, and crew that had gathered. He listened to the opinions of the archaeologists and took a close look at the specimen.

From the geological standpoint, Josenhans could see that the piece of rock was unusual. It was clearly different from the local "country" rock, which he had been dredging up and studying for so long. This chunk of stone was much more vitreous, much denser and more glasslike. (It proved to be a piece of basalt, a fine-grained igneous rock produced under intense heat in the cauldron of the earth's mantle.) That alone made it likely that the object, or the larger piece of stone that it was made from, had been intentionally carried to this area from somewhere else. The edges were also very sharp, which meant that it might have been used as a tool for cutting or scraping. But as for its other features, he could not be sure. "This is out of my field. I can't judge it," he told the others, and stood back quietly to wait.

At last, summoned back to the ship by radio, Fedje, the most experienced member of the archaeology team, arrived. McSporran handed him the piece of rock, and his jaw just dropped. Even a quick inspection told him what the others had already guessed. It had to be an artifact. Not only was it not made of local rock, but the shape and features—especially the pattern of blows required to shape it—indicated that it had been split off from a larger chunk of rock by human effort. Then its edges had been further worked and sharpened by systematic flaking, probably by hammering it with a punch made of bone.

Fedje had to assume that the piece had been left on shore when the sea level was low. It was conceivable, though, that it had been dropped into deep water from a canoe thousands of years later. This seemed highly unlikely, but could not be absolutely ruled out.

If Fedje's assumption was correct, based on the already well-known sea level record, the tool was at least 10,200 years old. This made it almost 1000 years older than any other artifact found in the Charlottes. It was also essentially equal in age to the bone artifact found by Tim Heaton two years earlier in Alaska.

The find lent extra credibility to the date of the Alaska artifact, and it showed that undersea archaeology by dredging was viable. Overnight, the Charlottes team was swamped with requests for media interviews. *U.S. News & World Report* used the event to lead off a long and detailed cover story, "America Before the Indians." *Scientific American* presented it as the first-ever proof "that people once lived on the now submerged lands." "It's great," Fedje told the *Seattle Times.* "We're finding this old landscape with all these interesting features—rivers and terraces and lakes—and we can show that people were actually using that landscape." Future access to ship time was assured. Next they would try in deeper water, where anything found would be even older.

CHAPTER FOURTEEN

Archaeology's Gold Standard

THE SHY BLOND teenager with braces on her teeth came up to Jim Dixon and handed him a very thin and delicate piece of dark glassy material a little over an inch long. Dixon put on his reading glasses to examine the perfect, unbroken microblade. "Very nice," he said. "Obsidian. Banded obsidian," a volcanic rock. He handed it back to Dylan Reitmeyer, his youngest volunteer digger. Her mother, Lynn, a Colorado schoolteacher, was also spending the summer at the cave, cooking for both the paleontology and archaeology crews.

I fished the closeup lens attachment out of my daypack and screwed it onto my camera. Dylan held the tiny artifact in the tips of her fingers while I snapped a few photos. Dixon put it into a tiny plastic envelope, filed it away in a box and suggested that we follow her to see where it had come from. She climbed down into the square hole that she had been slowly excavating for several weeks and pointed to a line of earth that ran along its wall near her feet. Dixon nodded. It was just what he'd expected. Given the depth, almost four feet below the surface just outside the cave, it was from the same period as the other microblades they'd been finding, about 9200 years ago. As the crew dug down and reached that layer in their individual holes, they were finding so many microblades that Dixon had begun calling that stratum the "microblade horizon." The early hunters and toolmakers who had left them there he dubbed the "microblade people."

Around us, laid out on the forest soil under the huge blue tarp, was a checkerboard of excavated and partially excavated squares, seven or eight in all. Each was defined by level red strings. Extending out from the cave mouth, and perpendicular to the cliff that rose above it, was a trench that had been dug right down to bedrock five or six feet below the surface. Each digger scraped away with a bricklayer's trowel and switched to fine brushes whenever anything interesting began to appear.

Meanwhile, Heaton's paleontology crew worked inside the cave. Each team was excavating its own carefully measured and well documented grid. Where the two grids met they called the Heaton-Dixon Line.

The archaeology crew was a mix. The older and more experienced crew members were receiving salaries. Others were student volunteers who were just trying their hands at archaeology for the first time.

Craig Lee, the crew foreman, was a slender soft-spoken guy with a beard and impish grin who was just completing his master's degree in archaeology at the University of Wyoming. This was his second season at On Your Knees Cave, and he was committed to working here the following summer as well, when Dixon would wrap up the dig. Lee was going on to a Ph.D. and planned to become a professional archaeologist. The dig was giving him not only experience but also a perfect research topic for his doctorate. "We've got a pretty good selection of obsidian from this site," he told me. "Visually distinctive types of obsidian. Basically my thesis is going to be to look at these different types using the X-ray fluorescence technique. I'll try to look at where this stuff is coming from—the different source areas—and then make some inferences about trade networks or at least population movements around this region."

Dixon already knew that the obsidian they were finding did not originate on Prince of Wales Island. The nearest possible source was Sumez Island, about thirty miles away. Over the next couple of years, Craig Lee would trace some of it to Mount Edziza, a volcano on the mainland around 120 miles away. People had been transporting precious stone long distances even 9200 years ago.

As crew foreman, Lee was responsible for most of the day-to-day work, especially when Dixon had to be away. It was also his job to show the new people, the volunteer students, what to do, but he played down the difficulty. Lee seemed to have found the right profession, because he loved camping out and found fieldwork to be mainly an adventure. Sometimes it gets tiring, he admitted. It was not all peaches and cream. And the weather in coastal Alaska, though beautiful the entire time that I was there, could be very wet. "By the end of the summer, you've spent eighty days in your tent, with mildew and everything else." Being away from home could also have its downside. Lee was not married, but there was a "significant other" in his life. "I think she'd be missing me," he laughed, "but she's actually off to an archaeology dig in Wales. So, I don't know. As of her last letter she was still missing me."

Another highly experienced guy with a special role was Eric Parrish. He was tall and gangly, with a beard and long black hair that he tied back to keep it out of his face. I was beginning to wonder about all these guys with beards. Did they wear them year-round, or only when they were out in the bush, where shaving was a bit of a hassle? Parrish worked full-time with Dixon at the Denver Museum as an artist with extensive knowledge of computer graphics. For Dixon's new book, Parrish had done all the maps, drawings of artifacts, and nice evocative sketches of how people would have used tools and weapons. At the dig, his tasks included measuring and sketching the locations of large artifacts as they came to light. He also did much of the screening of what was excavated, looking for smaller artifacts, charcoal, pieces of wood, or anything else that might help with dating or reconstructing the ancient environment.

Digging the grid with Craig Lee and Dylan Reitmeyer were two student volunteers, Heather Mrzlack and Heidi Manger. Both were here to learn and consider whether they might want to make archaeology a lifetime career. But Dixon emphasized that this dig was "not designed to be a place to train or teach people." It was not a field school. There were no lectures after the day's work ended, and no textbooks. Mrzlack, who had to prepare separate meals because she was the only vegetarian in a crew of voracious carnivores, had finished a B.A. in psychology but her interests had shifted. "I'd always been interested in anthropology," she told me, and had gone on an archaeological dig to Peru the previous year. Then she'd heard about Dixon. "It sounded like a really neat project." And she still thought so, even though it had its dicey moments. The previous day, a bear, attracted by the smell of shampoo, had ransacked her personal tent. She found the crunched and empty plastic bottle a short distance away in the woods. On the positive side, though, were the exciting finds. Her most impressive one was a sizable quartz crystal that must have been brought by people to the site. "It was big, a little smaller than fist-size. It was really cool."

She showed me her digging technique. In addition to the trowel and brushes, she kept a long, soft wooden stick close at hand. Whenever she found something like the quartz crystal, she would switch to using the stick. "You don't want to damage it with a trowel. And you want to get it *in situ*, so you just dig around it." As I watched, something white appeared along the dirt face. Mrzlack slowed down, picked around it carefully with the stick and then worked her way closer with the brushes. It was another quartz crystal, although in this case just a tiny pointed thing, translucent and mottled. Before removing it, she called over Eric Parrish and Craig Lee. Using a tape, she determined its distance in horizontally from the corners of her square. Finally, she recorded the distances in her field book, plucked the crystal from the wall of the excavation and handed it to Parrish, who put it into a plastic envelope and filed it away.

Working next to Mrzlack, and in an equally soiled T-shirt, was Heidi Manger. Like Mrzlack, she was in her twenties, slim and fit. Manger had recently graduated in English and Anthropology from a college in Denver and wanted to explore archaeology. This was her first fieldwork. "Craig showed me what to do. An hour's training," she laughed, "with a lot of questions in between."

As I watched, her tools uncovered something that looked promising. "It's probably a white chert flake," she said as she cleared away the blackish soil. The visible part of the object grew larger. "There have been a lot of these in this cultural layer," which was an easily recognizable stratum of gray and black clay. "They found five really nice flakes in a row right over here, and it's the same material as this." She pulled out the flake and cleaned it off to show me. "It's really rounded," she commented, "probably just from the process of toolmaking." It was likely a rejected piece of stone, what archaeologists call "debitage."

Still, she mapped it in by measuring from the level datum line down to the little gouged hole it came from, then out from the edge of the trench wall, and finally out from another edge of her square. She put it into a tiny plastic bag, assigned it a code number, and recorded everything in her field book.

Manger pointed to an adjacent rectilinear hole. "This square had some cool stuff, including two chunks of obsidian that were really well shaped. One was called—what was it?—a 'spokeshave.' It had this little nip for scraping and shaping." In fact, they had found two such tools. "And then there was a possible red ocher palette right here, four or five inches long, by three inches. That's what Jim was calling it. It was red and really large."

Dixon was puttering nearby and came over to give his take on the unexpected variety of materials the crew was finding. Quartz crystals, ocher and obsidian, all of them brought from quite a distance. "These exotic rock types show that people had already established trade networks along this coast, and were capable of intercoastal navigation." I mentioned that, in the Queen Charlottes, Daryl Fedje's group had found some telltale types of deepwater fish bones, such as halibut. These indicated that more than 9000 years ago people were fishing well offshore, which implied seaworthy boats. They were not just catching fish on shore in traps, or from the beach with some kind of net.

"Oh, sure," said Dixon. "It's the same thing here. At the time that this site was occupied, this peninsula where the cave is located was a separate island." Sea level had been higher than it is today. "So, unless this guy was an incredible swimmer, he had to come here by some kind of watercraft. And the fact that all this stuff is traded in here shows that these people had watercraft and were moving up and down this coast."

As Dixon excavated, his image of the site had acquired new dimensions. "At first I thought, maybe this was a bear-hunting locale, because before there were firearms it was easiest to hunt bears when they were hibernating. So this would have been a *hibernaculum*, a denning spot, for bears. But now it's clear that more than that was going on here. In that microblade horizon we've got big chunks of obsidian, big cobbles of it," which people obviously carried up this hill and well inland. "Also, someone has carried up a big quartz crystal, and they've flaked that," leaving a semicircle of debitage outside the cave.

"So they're bringing raw materials up here and actually making tools up here. Unlike at most hunting sites, where you have all your equipment ready, and you go in. You know, you don't sit outside in front of the bear den and make the tool and then go in. You come fully equipped. So, people were actually staying here for a period of time. We've also found a fair amount of charcoal, which suggests that people may have been here for camping, or cooking here. Also, just the other day we found a piece of red ocher. Hematite, that is. It's a pigment. You can rub it and make a powder, and it's commonly used as a basis for paint," possibly a body paint. "So, as we dig, we're learning more about the complexity of the site."

Working down among some big rocks in the entryway of the cave was Patrick Olsen, the only Native American intern currently on site. He was also the only member of the archaeology crew who preferred to camp at the beach.

A second native intern, Yarrow Vaara, had wrapped up her work a couple of weeks earlier. Vaara had been popular with the crew, in part because of her close links to the island community. The cave site was so remote that for weeks on end the diggers would see nobody from the outside world. And when they did get a day off for R & R, there was usually only time for a quick trip to the mainly nonnative outports of Port Protection or Point Baker. It had been a real highlight the previous summer when they were invited to a gala totem pole raising and feast at Vaara's home village. She had donned a traditional Tlingit button blanket, made of red woolen cloth appliquéd on dark blue with traditional family crests outlined in mother-of-pearl buttons, and had danced at that event. Just being with her made the others feel more welcome among the native population.

Patrick Olsen, though, was Haida. Short and solidly built, he had curly black hair and—wouldn't you know it—a beard. He liked to wear T-shirts with attitude, bearing messages like "Die Yuppie Scum." His father was from Kasaan village on Prince of Wales Island, but had moved away from the island to work and study at a college in Sitka. He died when Olsen was young, and Olsen himself grew up in the small city of Ketchikan. Although his higher education had all been down in the Lower Forty-Eight, he still liked coming back to the island for cultural events and deer hunting. He had a very fine rifle stashed in the drying tent down at the beach campsite, and was hoping to get in some hunting after the dig wrapped up.

That rifle was not the only gun around. There was also a shotgun in the cooking yurt, in case of a bear intrusion. And that's not all. Near the end of my visit, Dixon had to spend a couple of days attending to business in Craig, the island's main town. As he prepared to hike out, he strapped a huge Colt revolver to his hip.

When Dixon was out of earshot, I quipped to Olsen that it was like the scene from *Indiana Jones,* where the swashbuckling archaeologist cleans and loads his pistol. Olsen explained that Dixon once had a life-threatening bear encounter and never wanted to find himself in such a vulnerable situation again. Nor was Olsen himself taking any chances. "Didn't you notice this?" he asked, pulling a little black pouch around from the back of his belt. He opened the flap and cradled a very compact pistol. "It's a Glock automatic," he said, "a really nice gun."

It turned out that everyone else who had been hiking that trail each day was packing at least something to ward off bears. One had a can of mace-like bear spray. Another had an incredibly loud boater's air horn to scare them away. Only I had been naive enough to march up and down through those woods with trust in nothing but the odds. Now I was spooked. At this point, I was only going to be there for one more day. But on those last few hikes, I stuck close to Olsen and his handy little Glock.

Olsen had attended universities in Washington, Oregon, and Idaho, and after completing his master's degree he planned to go on to a Ph.D. in archaeology. He had done summertime heritage work—a cemetery cleanup—on Prince of Wales Island, and had demonstrated his interest and professional ambition to Dixon and Terry Fifield, which eventually won him an internship and place on the crew.

Olsen kept working in his shady spot halfway down into the entrance to the cave while we talked, an awkward position that forced him to reach in under some boulders and pick away at the soil with his trowel. I wondered aloud if this grueling routine would turn him off archaeology in the long run, but he just scoffed. In fact, he thought it was "kind of nice, a different way to spend time in the forest." Different, that is, from hunting. On further thought, he agreed that the work could sometimes become tedious. But that was true, he said, "no matter where you go in archaeology. And that's what you have to put up with. You have to do the hard work in order to find the interesting little pieces of information, which piece together the story of what people were doing here. And this is where you get it." He recalled his summers as a student in Ketchikan. "I used to work in a fish cannery, and that's a whole lot more mind-numbing than standing out here and being able to see the forest. When you're knee deep in fish guts, *that's* tedious. *This* is exciting," he laughed.

During the first season, the summer of 1997, Dixon and his team had dug the deep trench that ran out from the cave mouth. This gave them a good cross section of the deposits outside the cave. Along with quite a few artifacts, the trench material included charcoal from the microblade horizon, which was 9200 years old. By then, Dixon also had a laboratory analysis in hand for the human bones that Heaton had found. The maturity of the teeth in the mandible, or jawbone, and the size and shape of the pelvis, showed that the individual was apparently a man in his twenties. The relative proportion of carbon isotopes in the bones was also revealing. Carbon has two stable isotopes (carbon 12 and carbon 13) and one rare, radioactive isotope (carbon 14). The ratio between carbon 12 and carbon 13 in animal or human bones varies slightly with the diet that the animal or person has consumed. Study of the Alaska human bones showed that the person had consumed a diet extremely high in fish, sea mammals, and other marine organisms, as opposed to terrestrial animals and plants.

The fact that he had subsisted mainly on things like seals, shellfish, and seaweeds meant that an adustment in the radiocarbon dating was required. Initially, it had indicated an age of 9800 years. But as Dixon explained, it was well known from studies elsewhere that atmospheric carbon took quite a long time to become fully absorbed by the ocean environment. It could remain for hundreds of years, dissolved in the deep ocean, before working its way up the food chain to become part of marine organisms such as shellfish. This delay became known through discrepancies that had been observed repeatedly when scientists compared radiocarbon dates for marine shellfish and wood from the same deposits. The shellfish appeared to be substantially older than the wood. But that did not make sense; the trees and shellfish had likely lived at the same time. The discrepancy, they realized, arose because the wood had absorbed carbon from the air, while the shellfish had absorbed older carbon, which had been dissolved in ocean water for centuries. A similar discrepancy—and error—would therefore occur when dating the bones of any human who had lived mainly on shellfish and other marine organisms.

The phenomenon was known as the "marine reservoir" effect, and it required a correction to the dating. The exact amount could vary from region to region, and no one had worked it out yet for the waters around Prince of Wales Island, but Fedje and dating expert John Southon of the Lawrence Livermore laboratory had carried out the necessary comparisons and calculations for the Queen Charlottes. The correction factor was 600 years. The human mandible and pelvis, therefore, were really 9200 years old, just like the main occupation layer outside the cave.

My first thought was that Dixon must have been disappointed to learn that the human remains were not quite as old as initially believed. But the opposite was true. The concentrated marine diet supported his ideas about early coastal occupation.

There were critics, he said, who doubted that people were truly based in the coastal habitat so early, with a fully maritime way of life. At conferences, when Dixon presented evidence such as the early human remains and the bone artifact, these critics would object that such people might have gone out to the island for brief forays, that they were almost "picnicking." Dixon twisted his face in mock outrage at the idiocy of such an idea. "The conservatives would say, well, they were just interlopers, you know, just visiting. And that's clearly not the case." The marine diet was proof that whoever this person was, he had lived out on the coast for an extended period. "So when the isotope analysis came in, that was pretty conclusive."

"I want to return to the microblade horizon. I think it's important for the peopling of the Americas question, because we've got these exotic materials that were traded in, these different types of stone. We have evidence of the maritime adaptation, which comes directly from the human remains. And we have evidence of intercoastal navigation and watercraft. It's all inferential, except for the stone, which is pretty direct evidence. To come into a site and find these widely traded materials and this adaptation suggests to me that this isn't the first person." The first person, or small group of people, to begin hunting and gathering in a region would not be bringing with them such a variety of materials. "So this represents a much later cultural development, I think, out of a long period of occupation. It tells me that there's got to be older stuff here."

Of course, he already *had* older evidence, the bone artifact from the cave, which dated back 10,300 years. His confidence in that date was bolstered by the 10,200-year-old stone tool from the Charlottes. And even those dates did not likely represent the first people in the region. Dixon was convinced, based on the odds and the very limited resources that had been devoted so far to the coastal archaeological search, that people likely reached the area even earlier. How likely was it, he asked, that he and Fedje had found the oldest coastal settlements on the North Pacific? Not very. The chances were that much older sites existed. The problem was that they might well be underwater.

While the crew dug their individual squares and screened their material, Dixon was mainly busy perusing the printer's proofs of his latest book, which would be published in late 1999 as *Bones, Boats and Bison.* I had time to help with the proofreading and found some novel information and arguments. The book included overviews of all the key archaeological sites in the Americas that seemed relevant to the first peopling of the Americas. Dixon went point by point through the characteristics of each one, the artifacts or human remains found, and any problems with interpretation or dating.

Although he himself was a critic of the Clovis First idea, he painstakingly focused on the difficulties in accepting most sites for which pre-Clovis dates have been claimed. He explained why in most cases, according to the strictest standards of archaeology and geochronology, no one could be certain about the validity of the claims. In the end, Dixon, like most archaeologists, could only fully support one pre-Clovis site as well-dated and entirely solid in its interpretation. That was Dillehay's Monte Verde site. Several others, he thought, were borderline cases but probably valid. Despite some doubts about the interpretation and dating, he thought on balance that Gruhn and Bryan's 13,000-year-old Taima-Taima mastodon kill site in Venezuela was "persuasive." And he was impressed by a few more recently dug sites in North America, especially the Schaefer and Hebior sites in Wisconsin, which seemed to indicate mammoth hunting or scavenging around 12,500 years ago. He wrapped up his book with a short synthesis chapter summing up the often ambiguous facts. And he explained why coastal migration, along ancient shorelines and offshore refuges that are now deep under the sea, was the best way to account for the evidence to date.

But to me the unique part of his new book was a chapter reviewing all the ancient sites with *any* human remains that have come to light in North America. I was surprised to see just how few there are that date back more than 8000 years: only about two dozen on the entire continent. And it was striking that of those two dozen, only three fell into the Clovis age range of 10,800 years or older, with the oldest being the 11,115 year-old Anzick site in Montana. There were no securely dated and significantly older human remains in South America, either. (One example from eastern Brazil *may* be as much as 11,500 years old.) The scarcity of ancient human remains, and the sharp cutoff at the 11,000 to 11,500 year period, had implications for the presumed timing of any coastal migration. It shed strong doubt on claims for a long chronology in the New World.

Human remains are the true gold standard in archaeology. Whereas naturally broken or sharpened stone can sometimes be mistaken for artifacts, human bones and teeth are unquestionable proof of human presence. Often, with today's AMS radiocarbon techniques, they can be dated directly, without destroying them and with high confidence in the result. Stone artifacts can usually only be dated by their association with organic matter, such as charcoal, which introduces complications and uncertainty. In fact, many of the doubts surrounding allegedly pre-Clovis sites are precisely because of the possibly contaminated organic materials used to date stone artifacts. But as Dixon's book showed, ancient human remains in the Americas are exceedingly rare. Just why is not entirely clear, but coastal migration may be part of the explanation. Perhaps people were usually buried in ways that ensured rapid decay. Perhaps only very rare circumstances, such as accidental death in a limestone cave, as on Prince of Wales Island, favored preservation. Perhaps the ancient population was very small. And, of course, perhaps they were mainly living on the coast, which is now underwater.

In any case, there is no inherent reason why ancient human remains cannot be preserved and found, *if* they are there. A recent study of the middle Stone Age Neanderthal peoples shows that there are some 450 known cases of their remains found across Europe and the Middle East. And there are about 150 examples of *Homo erectus* remains, the bones of much earlier protohumans.

So how likely is it that people colonized the Americas well before the last few millennia of the glaciation, as the long chronology school has argued? If they did, shouldn't archaeologists have found at least some of their skeletons by now?

Supporters of Clovis First have always considered the absence of pre-Clovis human remains to be supportive of their viewpoint. Dixon had a different interpretation. He did not think there was an open and viable ice-free corridor in time to account for Clovis, and he didn't believe that the fluted projectile point tradition had moved in a north-to-south direction. Finally, he had accepted the pre-Clovis evidence from Monte Verde long before the mainstream of archaeology did so. The conclusion he drew from the conspicuous absence of pre-Clovis human remains was that people had reached the Americas somewhat earlier than Clovis, but had remained almost entirely on the coast for the first few thousand years. And since the areas of most likely coastal habitation are now all under the sea, we simply don't have a record of these earliest people. At least not yet.

As he explained it, "there's no well-dated human skeleton material in the Americas that's over 11,000 years. That's really it. So I would suggest to you, just think statistically." If there actually *were* people in the Americas well before the later stages of the last glaciation—say more than 30,000 years ago at Pedra Furada in Brazil's interior—why wouldn't they have spread out and settled in a variety of locations during the many thousands of years between then and 11,000 years ago? Most likely "over time you would find some human remains dating to these earlier periods." The total absence of such remains, combined with their appearance around 11,000 years ago "would suggest that the colonization of the Americas was probably a very late event. I think it favors a short chronology, as opposed to a long chronology."

How long ago did he think they first arrived? I wanted to know. Would it be 13,000, to accommodate the 12,500-year-old date at

Monte Verde? Would it be 14,000, which is about the time all the major offshore refuges on the B.C. coast were emergent from the sea and clear of ice? "Fifteen maybe," said Dixon. That would be long enough ago to account for Monte Verde, sites like Taima-Taima in Venezuela, and perhaps even some of the later dates from Meadowcroft Rockshelter in Pennsylvania, which he also thought might be valid.

But he did not want to be dogmatic. "You know, we used to think that maritime adaptations didn't occur until relatively late, at least in the New World, and possibly in Asia. And that people weren't really capable of getting out on the sea and doing much. We had very simplistic models about people's movements. And now, it's very important to keep our minds open to myriad possibilities. There may have been people living on the land bridge, people moving along the coast, possible transoceanic contacts. *And various combinations thereof,*" he intoned, feigning a pompous professorial voice. People could even have reached the Americas and died out for some reason. They could have been too few in numbers to establish a viable and expanding population, or an unknown disease might have decimated them. "It could have happened many times. We just don't know. The important thing is to keep our minds open to all these things and to find ways in which we can test them, reliably test them. And not just theorize."

Speaking of theorizing, I wondered what he thought of the arguments based on genetics and linguistics, which in so many cases seemed to support a long chronology. Dixon did not consider these approaches to be anywhere near as conclusive as discoveries from archaeology. He was diplomatic and called them "interesting," but hastened to add: "I think a lot of these techniques are flawed. They're not as rigorous as finding some physical, tangible evidence that's datable," the way archaeological and geological and biological evidence can be. "And so, they're noble attempts at approaching these problems from different directions. But, in my opinion, it's all pretty inconclusive." Still, as in his earlier book, the *Quest*, he was not absolutely dismissing all the radical alternatives to relatively late coastal migration around the North Pacific rim. "What's important when we make these statements is that we're just ruling out transoceanic migrations. And that's always a possibility. So there could be some of these events, really early. Who knows?"

Recently Dennis Stanford of the Smithsonian and Colorado stone tool expert Bruce Bradley had hypothesized transatlantic voyages around 18,000 years ago. They thought there might be a connection between Clovis in North America and the much earlier Solutrean tool tradition in southern Spain, which included large spear points and dated to around the peak of the last glaciation. "I don't know how serious they are about this, but they're looking at it," said Dixon. "What they're suggesting is that maybe people first came to North America via watercraft across the Atlantic. One possibility would be to skirt the margins of the northern ice, say hunting seals and that sort of thing. Another would be to just have a transatlantic voyage, straight across. So who knows? Maybe you have people coming from all different directions in different time periods, landing in different areas and adapting to different environments." But judging by his latest book, Dixon considered coastal migration around the North Pacific some 15,000 years ago to be by far the most likely scenario.

He was not very optimistic about finding direct evidence of this migration, such as traces of the actual boats or other watercraft themselves. "Yeah, I think the evidence is likely to remain inferential for a long time. I mean, it could happen. Somebody could find a paddle or a mast or thwart or something in a site." And then, perhaps that ancient wood could be radiocarbon dated. "But I think it's unlikely. If you had things like rafts, they're not going to sink," which was unfortunate. Wood that sank into the oxygen-free marine muds of the sea bottom might be preserved in recognizable form, such as a bunch of logs lashed together. More likely, though, "you'll just have debris. You might not even know what it's from. That's going to be a tough one."

This meant that the best archaeology could do was show the early presence of people based on evidence from caves like On Your Knees or from under the sea itself. He was very encouraged by the improvements in technology that had allowed Fedje and Josenhans to target occupation sites on the sea floor. And he was heartened by the changing intellectual atmosphere that was now much more favorable to the search for coastal evidence. It used to be so controversial that simply getting funding was difficult. Now that had changed.

Not that the battle had been entirely won. It was still a fight to have his views accepted. Some specialists still had doubts that people really lived on the coast so early. "I gave a paper a year or so ago to the Society for American Archaeology saying that we had a human on this island, we had the isotope data well dated. And this guy got up and basically critiqued it by saying, well, you know, this is just one individual and one case, and it doesn't show a pattern."

"So, what is proof?" he asked, knitting his brow in mock horror. "Do you need an overwhelming amount of proof? Do you need hundreds of human skeletons? With isotope data on their diets?" His answer was that you have to build a case and establish a pattern. "You start building it bit by bit." But he was outraged that people would continue to dismiss the entire notion of ancient people inhabiting these offshore islands. "At first blush, this looks very much like a pattern to me. I'd be very surprised if the person found here is atypical. The statistics of it are that you would find a typical person, not an atypical one." Nevertheless, to pin down the case for early coastal migration he'd like to find additional examples. "Sure, that's important. One site is always important, but you really need to develop a pattern. And some people won't be satisfied until a true pattern develops. Repetition. Finding it over and over again, like Clovis and Folsom, where there's no doubt. If you just keep finding more and more, and they all date to the same time period, it gets really convincing." That's going to take some time, he granted. "But I think it's a good time to do it. People's minds are a lot more open now."

As for where to look, Dixon thought that other caves in southeastern Alaska offered perhaps the highest likelihood, because they are above sea level and the preservation is good. And he had hopes for underwater searches like what was going on in the Charlottes, especially with the new technologies becoming available. But the task, he felt, was pretty daunting. "There are only about three or four of us in all of North America who are trying to do it," he lamented, "and it's very difficult. When you look at the vast geographic areas and the immensity of the problem, that's virtually nothing."

Still, he found reason for optimism. Ten years ago, he said, the search for early coastal habitation was too controversial to receive funding or serious attention, but now, as his experience and success showed, that was no longer the case. When archaeologists hear about possible coastal evidence now, they don't simply write it off as something that has to be based on an error or misunderstanding.

"The important thing is that, for the first time, people are seriously looking. And if you don't look, you're not going to find it, unless it's just by dumb luck."

CHAPTER FIFTEEN

All Alone Stone

TOM GREENE JR. tossed me a friendly nod over the top of his reading glasses as he looked up from the lab bench on the *Vector*. He beckoned me over. We hadn't seen each other since my week at Sgan Gwaii four years earlier, when he was one of the most enthusiastic Haida diggers on the totem pole restoration project. It was a year after the initial success by the Charlottes team, finding the 10,200-year-old artifact, and they were searching in deeper waters, where anything they found would be even older. I had just come on board the *Vector* to catch up on the latest phase in the undersea research. Unlike my earlier cruise with the ship, Daryl Fedje and Heiner Josenhans had the vessel for only three full days. It was now the morning of their second workday.

Greene looked like the cat who'd got the cream. He held out a tray of stuff he was poring over. It was full of pieces of shell from large clams, as well as some leaves and bits of wood. "These are all over 11,000 years old," he grinned. "And we've got artifacts, too. Found them yesterday." He pulled a small, labeled plastic bag from a box on the lab bench.

"Really?" I asked. *Already?* Had they scored on the very first day? "For sure," said Greene. "One hundred percent definite. I spotted this one myself. A nice flake about one inch long. It came from Station Ten, about 135 meters deep," or over 440 feet. He held up a small, dark, and irregular piece of stone. I wanted to believe him, but was uncertain about the alleged artifact. Josenhans was standing nearby, listening and looking slightly askance. How old would that make it? I asked him. "About 11,000 to 11,500, thereabouts, or older." An artifact that old would be a really dramatic find. It would equal or even trump Clovis and push back proven offshore occupation by a thousand years. But Josenhans quickly added that he and Fedje were by no means sure it was an artifact. "We'll have to see."

On this cruise, the geologists and archaeologists had to share the *Vector* with some fisheries scientists who were doing a survey of abalone stocks in the Charlottes. Along with many of the fisheries people, most of the geology and archaeology crew had to sleep in tents at a shore campsite and be ferried out to the ship each workday. But Parks Canada officials had informed me a month earlier that, for unexplained reasons that probably made sense only to the posterior-protective bureaucratic mind—what if I got hurt, for example?—they would not allow me to stay at that camp myself, or to hitch a ride with the ship to the work area. I had to see to my own logistics.

At first, it seemed that the only viable arrangement was to pay a local outfitter to run me down from Queen Charlotte City to Juan Perez Sound by speedboat, rent me a tent and a small boat for getting around, and then pick me up again a few days later. I could camp at any of several places on shore and rendezvous with the *Vector* each day. That would have been expensive, and I was uneasy about camping and boating alone in such a remote area. Fortunately a much better alternative turned up.

The successful undersea work in the Charlottes had attracted lots of media attention. If a breakthrough on the early peopling of the Americas were going to be made anywhere, Juan Perez Sound was one of the most likely spots. *National Geographic* magazine had been following the evolving research and contending theories and claims closely, and their parent National Geographic Society had funded Tim Heaton's work in Alaska. The magazine assigned a freelance writer and photographer to rendezvous with the *Vector*, just as I had planned to do. With millions of subscribers, the magazine had deep pockets. Their freelancers chartered a fifty-five-foot Charlottes-based private yacht, the *Anvil Cove*. It was the same boat that Fedje and Josenhans themselves had hired in 1993 for some relatively low budget surveying and coring work. The *National Geographic* team were kind enough to allow me to piggyback on their charter, kicking in only a modest contribution to the overall costs.

I took the jet from Vancouver to tiny Sandspit airport, which handled three or four commercial flights a day, and where everyone knew each other on a first-name basis. A driver with a ponytail loaded my things into his jitney bus, asked where I was from, and charged twelve bucks, including tax and the ferry fare across placid Skidegate Inlet, for the fifteen-mile trip to Queen Charlotte City. It felt great to be back on those wild outer shores. Bald eagles perched in tall trees and chittered away. Herons and oystercatchers patrolled the mud flats. Clouds scudded in off the Pacific and banked up against the steep, partially logged slopes that fringed the inlet. During the ferry crossing, I got glimpses of a few gray whales, their tails flashing in the distance.

To play it safe, I arrived a day early and checked into a bed and breakfast. It turned out that one of the other guests there, a nurse named Barb, was also going with us on the *Anvil Cove* as a private passenger. While we journalists were shuttling back and forth to the *Vector*, she would try a bit of sea kayaking.

The weather in Vancouver had been beautiful, but in the Charlottes it was raining. That night, a full storm hit the islands and knocked out the power. With no electric stove to use, the B & B's owner cooked breakfast with a propane torch. A couple from Italy used the phone and came back to the breakfast table despondent. They had scheduled a boat ride to one of the ancient Haida villages, but the tour company had canceled because of the weather. All morning the wind howled. Trees whipped wildly back and forth.

Would the *Anvil Cove* cruise be affected? I wondered. The *Na-*

tional Geographic writer, Mike Parfit, and his wife and assistant, Suzanne Chisholm, were flying up from Victoria in their private airplane, a four-seat Cessna. Parfit, an airplane buff based in rural Montana, flew almost everywhere on his own. By removing the plane's rear seats, they were even able to bring along a small inflatable Zodiac and outboard, which gave them enviable freedom. The photographer, Ken Garrett, was scheduled to arrive on the commercial jet.

I spent the rest of the morning wandering around the small outport in the rain, fretting about the weather and wondering whether my journalistic colleagues would make it. I went down to the large government dock to check out the *Anvil Cove*. It was a heavy-duty steel vessel, which was reassuring, given the conditions. It had a schooner rig, a large pilot house and a raised "poop" deck that gave it much more interior space than most sailboats its size.

Fortunately, the wind eased up by late afternoon. I phoned the hotel where the *National Geographic* people had reservations and was relieved to hear that they had arrived safely. Meanwhile, though, due to the weather and the tides, our departure time had to be delayed until the following morning. We would miss the *Vector*'s first day entirely.

By the time we sailed, our group had grown further. Phil Lambert, a biologist at the Royal B.C. Museum in Victoria and an expert on sea worms, had flown up late the night before. Since the *Vector* team would be bringing up grab samples from deep water, it was a perfect opportunity to add worms from Juan Perez Sound to his collection of specimens. Small, thin, and with an earnest demeanor that invited ribbing, Lambert took a lot of good-natured abuse about his fascination with worms.

The *National Geographic* team proved to be good companions. Mike Parfit looked to be in his late thirties, with an athletic build, shoulder length dark hair, and striking facial features. He was almost the spitting image of the actor David Carradine. Susan Chisholm was younger, at least as tall as her husband, and also wore her hair long. They had a laid-back, jocular manner, and in their spiffy new matching parkas and rain gear, they could have been a poster couple for L.L. Bean.

Ken Garrett, the photographer, was a citified no-nonsense guy, single-minded about protecting his top-of-the-line cameras and lenses and determined to get those two or three key shots that would justify the long trip from his home near Washington, D.C. Garrett specialized in archaeology and paleontology and had been to Monte Verde, East Africa, and almost every other place in the world where the story of early peoples was unfolding. He was also extremely well read on the background literature.

The weather had turned clear and sunny, with early summer snow still visible on the higher peaks, when we packed our huge pile of belongings onto the *Anvil Cove.* Large as the boat was, we were crowded. The owners, Barb and Keith Rowsell, had brought along two young women to do the cooking and cleaning up. All the bunks would be full, with none for Phil Lambert, who was expected to sleep onshore at the scientists' camp. This was anything but a luxury cruise. It was sleeping bags, not linens. Simple food, not fancy cuisine. No beer or wine with the meals. Barb Rowsell had warned me to take a morning shower at the B & B. There would not be enough fresh water onboard. A sponge bath was all we could hope for.

She was slender, with long prematurely white hair, and was such a good organizer—a real soccer mom type—that she could direct the loading of the boat without interrupting her droll tales of eccentric Charlottes neighbors. Like her husband, she was an experienced and licensed charter skipper. In fact, she took the helm for most of the trip south through the islands along the eastern side of the archipelago. Now in their forties and fifties respectively, the Rowsells had begun raising their family while living at an extremely isolated homestead on the wind-battered west coast of Vancouver Island, where they were practically the only nonnatives and everyone relied on generators and radio phones. By comparison, Queen Charlotte City was the lap of luxury and convenience. Keith Rowsell was a quietly and serenely competent guy with gray hair and—you guessed it—a beard. By trade, he was a machinist and skilled welder who at various times had earned an excellent living working at logging camps. But the Rowsells also loved the sea, and Keith, whose friends all called him "Flash," though he was anything but flashy, had built the *Anvil Cove* himself.

Our route first took us out into Hecate Strait and around the eastern side of large Louise Island. We passed Skedans (also called K'una), a haunted abandoned village that I had visited more than a decade earlier with my friends Steve and Marlene during the shakedown cruise of their yacht. We had walked into the dense woods and stared in silent awe at the few remaining totem poles. The rest had either been removed to museums or had fallen over and been engulfed by the ferns, mosses, and salal. So had the frames of the longhouses, where almost 500 people had once lived.

Because we would be coming through the area, Flash had agreed to drop off supplies at a camp on small Limestone Island, where a group of scientists were studying the nesting behavior of the ancient murrelet. We had a few cartons of groceries for them onboard. Lashed on deck was a drum of gasoline. "We all help each other out here," said Flash. The charter boats and the outfitters, the flying services, the fishermen, the Parks people, the Haida. "If anyone needs a delivery, and someone else is going past that area, they'll drop it off. Or they'll pick someone up and take them back to town. Whatever." At Limestone Island, one of the researchers came out to meet us in a Zodiac. The Rowsells handed him down the groceries. Then the drum of fuel went over the side with a big splash. As we moved on, the guy was tying a rope around the drum to tow it to shore.

A few miles farther and we were cruising past Richardson Island, where some of Daryl Fedje's colleagues had continued excavating after the five-year inventory survey ended. It had become the most intensively studied onshore site in the region. Later that summer, biologist Rolf Mathewes would be coring an interior lake on Richardson to determine exactly when the local glaciers retreated.

At last, late in the afternoon, we entered Juan Perez Sound, a fifteen-mile-long expanse of water that was partially open to Hecate Strait but sheltered from the worst of the prevailing southeasterlies by several islands. I ducked into the pilot house to look at the chart and orient myself. On the northeast corner of the sound was popular Hot Springs Island, where kayakers loved to loll in warm pools overlooking the sea. At the southern end was narrow Burnaby Strait, out of which glaciers once ground their way northward into the sound. Along the western side was Matheson Inlet, a deep slash into mountainous Moresby Island. The inlet opened out into a crescent-shaped part of the sound called Werner Bay, where the 10,200-year-old artifact had been dredged from the bottom the year before. Smack in the middle of the sound, with no other land nearby, was a strange, humpy little islet called "All Alone Stone." Fully living up to its name, it was at best an acre in size and had only about a dozen trees clinging precariously to it. While still miles away, we could see through binoculars that the *Vector* was working close in to the islet. The early evening light was turning soft and warm, just the kind photographers love. Ken Garrett was excited. One "must-have" photo, he figured, was a distant "location" shot of the ship with striking background scenery. At the speed we were going, it would take a while to reach the *Vector*. Garrett readied his cameras and planned to capture the ship dredging. The ship and bizarre islet would be backlit by a rapidly reddening sun that hung low over dark green mountains fringed with white frosting. A spectacular image.

Alas, it was not to be. Just as we were close enough to make out the white Coast Guard slash painted on *Vector*'s red bow, she began to move off. The long workday was over, and she headed for an anchorage behind Huxley Island, where the campsite was also located. We followed, anchored nearby and sat down to dinner. Afterward, Parfit and Chisholm inflated and launched their little Zodiac and took it for a short spin around the anchorage. Meanwhile Flash took Phil Lambert, our worm biologist, ashore in the *Anvil Cove*'s big Zodiac to join the group camped on Huxley Island. But soon they were back. The campsite was packed, and no one had an extra tent for Lambert. "Oh, no. I guess we're stuck with him," we chimed in on cue. There was a foamy onboard, so he slept on the cabin floor.

* * *

The next morning, a speedboat fetched Lambert, and the *Vector* headed out to spend the day taking grabs and some cores around All Alone Stone. We soon followed. The plan for the day was that Flash would hover around the research ship in the *Anvil Cove*, staying near enough for the *National Geographic* people to come back and forth easily in their small Zodiac anytime they wanted, including perhaps for lunch. Garrett could use the small Zodiac's mobility to get shots of the *Vector* at work from any direction or distance he wanted. Barb Rowsell would go off kayaking with Barb the nurse. And I would spend all day on the *Vector*.

The large crew of geologists and archaeologists were already hard at it by the time we reached the ship. The *National Geographic* team, who had never been on the *Vector* before, wandered off to look around. I was greeted on deck by a couple of the people I already knew. The ship's cook was out on deck, too. Noticing our group of newcomers, he buttonholed me to ask who I was and whether or not I planned to stay for lunch. I said, Yes, if possible. He asked about the others—Parfit, Chisholm, and Garrett were down at the far end of the deck—indicating that the ship was packed to overflowing and he already had more mouths to feed than planned for. I said that he might want to ask them about their exact plans. But I made the mistake of adding that since they had the small Zodiac at their disposal, they could always zip back over to the *Anvil Cove* for lunch. The cook looked relieved and mentally crossed them off his lunch list. This would come back to haunt me.

Walking around the ship, I saw that things were quite different from what I had witnessed on the 1994 cruise. The focus then had been almost entirely on marine geology (surveying with the sidescan and bottom profiler, and taking sediment cores) with only a few hours for dredging with the grab. Now everything was geared up for an intensive archaeological effort: dredging and screening. There were three screening stations set up along the starboard rail, each with a high-pressure water hose. The screeners were Fedje, archaeologist Quentin Mackie of the University of Victoria, and Tom Greene.

On the aft deck I could see that the dredging was almost a production line operation. The *Vector* had undergone an official change of administration and a new paint job to go with it. Until the mid-1990s the ship had been run by Canada's Department of Fisheries and Oceans, which was a fairly relaxed outfit. There was not much of a dress code, and although the crew worked long hours, they also took their time about their tasks. Now the ship was part of the Canadian Coast Guard. Although not a military service, it had a certain dedication to spit and polish, clean uniforms, and a gung-ho, snap-to-it work ethic.

Presiding over the deck crew was a bosun named Jamie Henderson, a baby-faced giant of a man who had fashioned them into a crack team. He held a walkie-talkie in one hand to communicate with the ship's bridge and gave hand signals with the other to the four or five deckhands. They operated the winches and hydraulic A-frame that deployed and retrieved the grab jaws.

Five years earlier, it had taken about half an hour to send down a grab, bring it back up, dump the contents on deck, and shovel it all up into boxes and buckets for screening. Now the deck crew were doing it in as little as ten minutes, and they had the ship moving on to the next target location before the grab was all the way to the surface. In fact, the crew moved the sludge so fast that the three screeners could not keep up. After an hour or so, there were so many bins of bottom dredgings lined up on the side deck, waiting to be processed, that there was time for the ship to take a few cores as well.

"We're really pounding away at those terraces," said Josenhans with a satisfied smile. These were the places along an ancient drowned drainage system where he and Fedje thought people would have concentrated late in the glaciation. Josenhans showed me more of the encouraging stuff they had brought up the previous day: chunks of wood, clumps of peat, blackened but still quite identifiable leaves and broken clamshells. All from a sea bottom that was last exposed to the air 12,000 years ago. The wood and leaves showed that there had been forest here. The following day they found even more striking proof: the stump of a small pine tree *in situ*, with its roots still buried in the mass of intact material brought up by the jaws. He was especially happy to see the great density of clamshells. Shells dredged the previous year had demonstrated decisively that there were ample seafood resources here at least 10,000 years ago. Now the same was obviously true back at least to 12,000 years. Amazingly, the shells hardly showed any decay. Some looked as though they had just been harvested.

What Josenhans was most eager to show me, though, was the new technology he now had at his fingertips. I had heard and read about it, and I had watched his video years before, but I had never seen the swath bathymetry and sophisticated computer program in action. "Come over here," he said, leading me to a desktop computer at one end of the lab and introducing me to Jennifer Harding, the graphics techie he'd brought along from his home institute in Nova Scotia. He pointed to a large-scale image of the entire working area on the monitor. "Yesterday, we were working here, north of All Alone Stone. And we would look at the seismics," the contoured false color maps of the bottom that were so much like three-dimensional aerial photos. He and Fedje would then decide exactly where to put down the grab. "We pick the sites based on the morphology and our interpretation of the morphological development," in other words, the geological processes that created the now-drowned landscape late in the glaciation.

"We can see, for example, how the old river cut down through here and developed this terrace edge." Even I had no trouble recognizing the river channel and flat terraces. It was this extremely detailed information about the seabed that let him zero in on the best places to deploy the dredging jaws. "We're actually going to pick a couple of targets now, Tom. You can have a look." Harding worked the keyboard.

"This is All Alone Stone, here," said Josenhans. It looked like a little mountain, peaked in the middle with squiggly contour lines running around its flanks. "This is the meandering old river system flowing out to the delta, which is right here." He pointed to some lines on the screen that splayed out like the fingers of a hand. "And the really interesting targets that we've had are in this area here, which are south-facing old terraces, protected by high bedrock ridges here." He ran his finger over a series of parallel, raised features. "What we did was, we zoomed in on this target here. You can see it on this blowup. Can we show the image of where we were yesterday, Jen?" Harding clicked her mouse and everything got much larger. "So these are the precise positions—see the little exes?—where we actually sampled yesterday," said Josenhans. "And the most remarkable site was here." Josenhans pointed to Station 8, smack in the middle of the terrace.

Fedje had mosied by. "Those two grabs had a lot of wood in them," he chimed in.

"We're finding wood bits virtually throughout," Josenhans agreed. "But the one that is really, really dense—it has a beautiful peat surface and wood embedded in the peat—is Station 8. Depth 142 meters. We can show you examples of the wood." We moved over to a nearby workbench. "Here's a leaf that actually came out of Station 8," said Josenhans, pulling it out of a plastic baggie. It was from a willow and was quite blackened. "It's a bit fragile, but still intact. This is typical—black, peaty, woody material," he added, indicating some moist, spongy looking stuff in a bag. "Daryl has his microscope onboard with him." Fedje and Josenhans had stayed up late the previous night inspecting the material from Station 8. It turned out to be freshwater peat with some marine diatoms that had been washed into it.

Next Josenhans opened a larger bag with a big chunk of wet wood. "This is typical of what we're getting down there." Wasn't it possible, I asked, that the wood fell into the water thousands of years later and sank? "That's a fair question," he granted, but then went on to explain how the stratification of the bottom sample proved that the wood was ancient. "There's a completely different layer on the surface. These are subsurface. The grab bites in about sixty centimeters," or some two feet. "And that sample there shows pieces of wood in the peat context. They had been buried by peat. So that's completely unambiguous. That's an *in situ* sample." And yet it was retrieved from some 463 feet beneath the sea, "which is essentially confirmation of what we've been saying about the history of sea level in this area. But the beauty of it is that we can now—through the peat analysis—get a really good insight into what grew here at the time. Because peat preserves the pollen in the best way possible. I mean, the leaves help. But looking at the pollen, we can establish exactly what grew here."

Fedje added that there were also pine cones and needles in the sample, which indicated the type of trees that replaced the earlier tundra vegetation and became a dense forest around 12,000 years ago. (Radiocarbon dating of the *in situ* pine tree stump found the following day showed it to be 12,200 years old.) "And the other part of it is, look at the shells," Josenhans went on. "These shells are intertidal, littleneck clams. They look modern, but they're not. We dated a number of these last year, and they came in right on target." They were exactly the age that the sea level history predicted. "And the neat part is that they indicate that there was a food source for marine-based people. There was an *ample* food source. Within the lagoon sediments there was up to 20 percent by volume of clamshells." There looked to be a similar volume here in the deeper water near All Alone Stone.

Back at the computer, Josenhans asked Harding to demonstrate a marvelous feature of her program. With a click of the mouse she added an imaginary sea to the map, a darkened shading with sharp, clear edges that followed the contour lines and indicated exact sea level. Then, by scrolling, she made that sea level rise and fall instantly, which showed what the shoreline would have looked like at any time in the past. Up and down it went, and the shape of the shoreline changed just as quickly. She enlarged the area showing the little islet where the ship was working and adjusted the virtual sea level until it showed just why this would have been an ideal dwelling site 12,000 years ago. I could see the old stream drainage, or "paleo river," as Josenhans was calling it, flowing from higher ground down to the sea. Just as Juan Perez Sound was now largely sheltered from the prevailing southeasterly winds and swell by islands in that direction, so would this smaller area have been at that ancient time.

Josenhans asked Harding to show me another feature. Once again, with only a few deft moves of the mouse, she could place an imaginary sun at any height she wanted in the sky and move it around the horizon. Doing so created virtual shadows and more subtle shadings on the landscape, and these were very revealing. All kinds of surface features stood out much more dramatically than they had from their colors and contour lines alone. Distinctive formations that geologists are trained to recognize appeared in a way that was never possible with the images from the sidescan sonar. One example was the elongated hillocks, or drumlins, that had been left by the glaciers. "It's a tremendously powerful tool for interpreting how the landscape evolved," said Josenhans, "based on its geological shape."

And there was still another level of sophistication. "The technology," Josenhans grinned, "allows you to fly through the image," akin to the way the imaginary camera did in the video that he had lent me. "We have a computer system called the 'Fledermaus,' which is the German word for 'bat.' You have a joystick and the computer in front of you. And it's a little bit like a cross between a helicopter and an F-18. You can hover, or you can zoom in at a great rate." He loved it. It was the closest thing to being an underwater pilot. (Josenhans was in fact a licensed pilot, although he did not own a plane.)

It was very different from trying to piece together a picture of the geology from the fuzzy black-and-white shadowgrams generated by the earlier technology. "That was state of the art. But it took more time and effort. This is a whole new dimension in terms of understanding. You can interpret a lot of the *process* that goes on in terms of sediment transport," which was geology talk for how a landscape is created dynamically over time. "And as a geomorphologist, that's the game, to interpret the origins of the land forms."

"This little feature here. See? That's a slump." Josenhans pointed to a distinctive feature on one of the slopes of All Alone Stone. "What happened there is that some sediment failed, sediment from this little basin here." He indicated a small and relatively level spot on the slope. "Evidently that material failed and slid down. This is a new level of resolution that we didn't have before. You can see it all so clearly. The sharp cliff face, and then down here, the river terraces." Using the detailed image on the screen, he ran me through the entire scenario of creation and destruction over thousands of years. "Let's go back and see how this landscape evolved around All Alone Stone." Late in the Ice Age, perhaps 14,000 years ago, sea level was already much lower than it is today, and there were local glaciers. "The glaciers pumped out a lot of sediment, and it filled this embayment here with outwash debris. Then the sea level fell further, and the river cut down through that outwash terrace to a new level. In fact, it did it in two increments. First it found this level here. You see that terrace there?" He pointed to a flat area. "And then it cut down even further, and it bypassed that terrace. So there's a little abandoned terrace—there—and it flowed out through this gulch here." The river skirted the very edge of All Alone Stone. "It was an extremely narrow gulch, which in terms of human geography is a very interesting place.

"So now we can go back again to the bigger picture. In terms of the overall scenario, we have this bedrock high, here, protecting everything against the waves from the east." He pointed to a raised, elongated hump of bedrock that looped halfway around the area. "We have this sentry post here—All Alone Stone—if you think of it in terms of human defenses. We have this lagoon environment here," in behind the bedrock high. "And we have a southwest-facing terrace on former peat, which is an interesting place for people to have lived. And that's where we're sampling right now."

I went back out on deck to watch for a while as the dredging jaws kept coming up like clockwork. Then it was lunchtime. I tagged along with Phil Lambert to the mess. There was a choice of main dishes, and I ordered the "Baron of Beef *au jus.*" This sounded a little grandiose for a meat sandwich on a long bun, with a little bowl of beef broth for dipping. Actually, though, it was pretty good. Later I headed up to the second deck to take a few pictures of the action below. There I ran into my *National Geographic* colleagues, who were sitting in a sheltered spot and eating sandwiches.

Because the cook had seen fit to exclude them, but not me, they had taken the small Zodiac back to the *Anvil Cove*, prepared themselves some brown bag lunches and brought them back to the *Vector.* "How was lunch?" Mike Parfit asked me with a toothy, sardonic grin. *Uh-oh*, I thought, but I wasn't going to let them get the upper hand. "Not bad," I said. "I had a Baron of Beef sandwich." Parfit looked at his wife, who glanced down at the cold sandwich of cheese and sprouts she was just finishing off. They both looked over at Ken Garrett, and then all of them turned and gave me a withering stare. "Oh, Baron of Beef?" said Parfit in a "la-dee-dah" tone. "Baron of Beef *au jus,"* I added, just to twist the knife a bit. "Ah! *Au jus?"* Parfit echoed, with a faraway gleam in his eye. "Well, we'll just have to see whether there's room in the Zodiac for Mr. Baron of Beef, later, when it's time to go back to the *Anvil Cove."*

Through the afternoon, dredging went on apace. After each grab sample the captain quickly repositioned the ship and the jaws went down again. The action was fast and mainly routine. As Josenhans had said, they were just pounding away at those terraces along the drowned river drainage, hoping to find another needle in the haystack. I was in the lab when Lambert came in and told Josenhans and Fedje to come outside. Something interesting had come up. We all hurried out onto the aft deck and clustered around the jaws. Garrett elbowed his way through the little throng with his camera poised. If this was the big moment when something historic was found, nothing was going to get in the way of *National Geographic.* But Fedje told us to relax. It was just some more pieces of dark rock with sharp edges. They were not obviously artifacts.

Back in the lab, I asked Fedje what he thought of the pieces of stone found the previous day, which Tom Greene was so sure were artifacts. "Well, I don't know," he answered cautiously. "We're waiting to see. You know, we found lots of rock and things." But they were not really convincing.

Some of what they had found came from a very likely place for people to be camping, he admitted. "It's a really nice terrace, and it sits along this river channel. And on this terrace we found a lot of wood, from trees, that had been sitting on the edge of that bog. And there were a few pieces of stone that . . ." It was obviously tempting for him to take the plunge and say that he thought he might have found artifacts. But he wouldn't do it. "We'll wait and see."

I probed a little further. Did any of the stone objects look like they had been worked? "Yeah, we're not absolutely positive," he said, "but there's one piece that looks like it's what we call a cobble chopper, which is a large piece of granite material that's been split in half, and there's some chips off it. But you know, it's the kind of thing where you want . . . for this kind of situation . . . you want to be absolutely sure. So, if that's all we get, we're not going to say very much about it at all." There would be no announcements to the press, or claims about possible artifacts in scientific papers. He and Josenhans would stick to the positive environmental indicators and leave it at that.

I asked how different these sharp pieces of stone were from the stone knife found the previous year. "Well," he said, pointing to a poster on the wall of the lab showing the 1998 artifact and its location, "this one here is very clear, very obvious. And these other ones . . . we'll see. We have to find something that's absolutely . . ." He didn't quite know how to put it. "If it was an ordinary place, like an ordinary archaeological site," which was already known to contain artifacts, "we wouldn't have a problem with the stuff we've found. But because this is going to have to go through so much rigor in terms of scrutiny, we're not going to say that we're sure."

I wanted to know how he could be so sure about the previous year's stone tool. "Well, this is of a very fine-grained volcanic material. It's very glassy. It's the kind of rock that people used in the past, because it fractures in a very controlled manner. It's got very sharp edges. It came out of a rounded fluvial gravel bed of the old river channel. It has characteristics that show that the flake itself was removed using a particular technique." He described how such an artifact was made by applying soft hammering blows to the rock using a piece of antler or very hard wood. "That diffuses the force of chipping of the stone. It's a very characteristic technique."

Could he recognize with the naked eye that it was an artifact? "Oh, yeah, absolutely. It's a large artifact, about ten centimeters long," or some four inches. "It's got sharp, knifelike edges, and they're very well preserved. And you can actually see the marks from where people used it. It may have been touched up a little bit, but basically the primary edge was used, and you can see the nibbling from that use on the sharp edges of the artifact." Other than this edge use, though, "it's pristine. The artifact had sharp, very sharp edges. And it's in the middle of a gravel bed where everything else is basically round, a very high energy delta," with a lot of tumbling and grinding from the action of waves and flowing water when the delta was being formed. "And here's this stone knife, sitting proud in that material. So it couldn't have been transported very far, if it moved around at all."

We talked about the Monte Verde dig in Chile and the incredible, almost microscopic, examination that Tom Dillehay's results and claims were subjected to before his site was accepted. Was that what Fedje meant about scrutiny being a consideration? What if this were *not* a situation where finding a 12,000-year-old artifact under the sea would overthrow an entire scientific paradigm and decisively support a competing one?

Again, Fedje was cautious about assessing these possible, perhaps even likely, artifacts. "Well, anywhere else, we would say there's no question at all. But there are a few diagnostic features you need to have in order to be absolutely, absolutely sure. And so we'll wait. We don't find that, then we can't say anything for sure yet." For one thing, the pieces being dredged up that day seemed to be of native rock. That was unlike the previous year's artifact, which was made of rock that had to have come from at least thirty miles away, if not farther.

And so, the day's research ended inconclusively. They had found ample evidence of extensive food resources back 12,000 and more years ago. This was certainly a viable late Ice Age refuge for people. *If* people had been there. As for artifacts, though, all they could do was to keep putting down the grab and screen carefully. After all, they did not find the 1998 artifact until late in the third day out of four.

I headed back with the others to the *Anvil Cove* for the evening. The weather remained beautiful, enhanced by a lovely sunset over the mountains of Gwaii Haanas to the west. Mike Parfit and Susanne Chisholm took a cruise in their Zodiac and came back beaming that they'd seen a humpback whale. But if I thought my lunchtime betrayal had been forgotten, I was wrong. When I went to sit down at the dinner table, Chisholm pretended at first that she was not going to move over for me. "No room for Mr. Baron of Beef," she said. "Baron of Beef, *au jus*," Parfit chimed in.

The next day, the *Vector* was back on station early, dredging and also taking cores when possible. Once again, there were flurries of excitement when something interesting came up in the grab jaws. As on the previous day, we all rushed out and crowded around. But again it was the kind of stuff that only a geologist could love: "transgressive intertidal," Josenhans announced. There were bits of angular gravel with sharp edges that just possibly could have been used to cut or scrape something. But there were no obvious artifacts. "More of these definite maybes," Fedje quipped.

Quentin Mackie, the University of Victoria archaeologist, took a similar line to Fedje's about the "possible" artifacts. While he screened, he picked up pieces of stone to show me, and explained what he and Fedje would want to see before they declared something to be an artifact. "A number of these I would have no hesitation in calling artifacts if they came from a known archaeological site where I was digging, say on shore. Because there's always a gray area when you're looking at stone tools. There's 90 percent that you can tell for sure, and then there's 10 percent that you can't really tell for sure." If you call those artifacts, "then half of them are probably right and half are just natural events. Well, all we're getting here so far are those that fall in the 10 percent."

There were many ways a piece of rock could get such sharp edges. "It could be from freezing and thawing," he said. "Or it could just be gravel tumbling down. Or it could have been block-fractured, sort of tectonically," deep in the earth, and later came to the surface. "Normally," he added, " you'd only expect to find that occurring in one or two places on a piece of stone. But when you get pieces which have multiple fractures, which some of these seem to do, that means it got impacted more than once in its life. Then your law of averages goes down." That's why some of the "maybes" were interesting and would be looked at more closely. But for now they were still only maybes.

"But the thing is, with stone tools there are a set of distinct diagnostic traits that you look for. You look for concoidal fracture, which is a sort of wavelike fracture through the rock, which is only really caused by striking it a certain way. Basically it's a ripple. When you strike the right kind of very fine-grained material very hard, it sends a shock wave through it. It travels as a wave. And as the wave comes out the other side, the piece of stone detaches and falls off, but you can see the shape of the wave."

Couldn't the action of a glacier do that, I wondered. "Well, tumbling might do it if it was a very abrupt, sharp strike. But, typically you don't expect to see very many of those occur naturally just from rocks rolling downhill. But you expect to see some. So what you look for then are the signs of greater deliberation, i.e., two or three of those strikes on a single rock, which is very unlikely to happen by chance. We're not finding that here."

By comparison, "the one they found last year, there's no question. It has flake scars, classic flake scars. I don't think there's an archaeologist in the world who would dispute that that was a flake. Because you just don't get that kind of degree of extremely sharp, directional percussion flake scars on both sides—concussion, concoidal fracture, et cetera. You look at both sides and you see that there have been flakes detached from the other side, and that *it*, the flake itself, was detached. It's not just broken rock."

Another thing Mackie had to consider was the sheer odds. "I've pulled out a dozen of these pieces to look at, but there's probably hundreds in the screen, which are also broken, and so here I'm just creaming off the 1 percent of the broken rock that kind of looks suspicious. And that's the kind of percentage you might expect to get through natural events anyway. So, as I say, in the absence of the real honest to goodness definite artifact, just looking at these—no matter how many of them you find—it's not going to really do it."

We headed back to Sandspit with mixed feelings. Naturally, we'd hoped to witness a dramatic success. Instead, we shared a sense of the frustration that scientists routinely feel when their results are inconclusive. The environmental indicators (wood, peat, and abundant shells over 12,000 years old) were solid confirmation of a resource-rich offshore refuge. And possible—even likely—artifacts were encouraging. But they were by no means conclusive, and the search would have to continue.

CHAPTER SIXTEEN

Arlington Woman

THE ARTIFACTS FROM offshore islands in Alaska and British Columbia showed that people had been on the remote outer North Pacific coast by 10,200 or 10,300 years ago. That pushed back known coastal occupation significantly, but it was still 1000 years later than the earliest Clovis spear points. This time gap vanished when an old archaeological find from an island off Santa Barbara, California, was subjected to renewed scrutiny and dating.

In September 2000, I flew to Los Angeles and joined U.S. National Park Service archaeologist Don Morris for breakfast at the little airport near Camarillo. Santa Rosa Island, 63,000 acres of former ranch land, was now part of Channel Islands National Park. Morris had been responsible for archaeology there since the mid-1980s. Although it was a Saturday and should have been his day off, he had agreed to fly out to Santa Rosa and drive me halfway across the island to see the site at Arlington Canyon. Its sporadic excavation was still under way, and no one was allowed to visit without supervision.

Morris was in his early sixties, with thick, graying hair. Born in Texas, he had studied and worked for many years in Arizona and had an easy southwestern drawl and manner. He also had a slight paunch that filled out his gray Park Service uniform, and a big appetite. Which was fine, because our flight with Channel Islands Aviation was delayed until the early morning coastal fog burned off.

As we dug into our eggs and hash browns, I leafed through the Park Service handout that Morris had brought along. It included maps and a thumbnail history of the five major islands that comprised the park. Large areas on some islands were still in private hands, and the archipelago offered camping, boating, and scuba diving. There was also a brochure from the Santa Barbara Museum of Natural History, which had played the leading role in archaeology on Santa Rosa Island for decades. It featured a photo of the museum's most captivating recent find, a pygmy (or dwarf) mammoth elephant that was so well preserved it looked alive.

Morris knew I was interested in the park's archaeology because of how it fit the emerging picture of early coastal migration. So was he. In his many years working there he had given lots of thought to the search for traces of ancient coastal people.

At the peak of glaciation, 18,000 to 20,000 years ago, he told me, when sea level was low, all the Channel Islands were linked. They added up to an island, called Santarosae, that would have been about ten times the size of Santa Rosa today. "And by 13,000 years ago, they still would have been connected." This large Ice Age island was separated from the mainland by at least five miles of water. People would have needed boats to get there.

Morris thought there was great potential for underwater archaeology in the Channel Islands. He pointed to the map. "This area off the east end of Santa Rosa—it's on the lee side. What makes me want to look there is that there's a good possibility that occupation sites would have been covered by aeolian deposits," or windblown soil carried long distances by the prevailing northwesterlies. "So when the rising sea covered them, they would have had a protective 'hat' over them. We'd have a good chance of getting down there and finding an undisturbed site." He assumed that early people would have lived right on the shore. "That's where we live today. That's where people lived in the more recent period. There's a tremendous concentration of archaeology on the coast. So if we want to find what life was like 11,000 years ago, we have to find the coastline of 11,000 years ago, where it's still intact, and investigate there."

But Morris himself would not be the one overseeing any such underwater digs. He was counting down the last few months of his career with the Park Service.

Around ten o'clock the coastal fog burned off. Together with a family of day-trippers, we were shepherded out to our plane. It was a nine-seat British-built Pilatus Islander, a twin engine high-wing craft that Morris had flown on often and admired for its safety record and short takeoff and landing capabilities. This was reassuring, since we were flying out to a gravel airstrip. Almost as soon as our plane cleared the coast, we could make out the hilltops of long, snakelike Anacapa Island, where they jutted through a cloud bank on our left. A few minutes later we were over Santa Cruz, easily the largest of the Channel Islands. Parched and craggy, it had sparse vegetation, eroded canyons and deeply gullied badlands. One mountain reached almost 2500 feet. Morris shouted over the engine noise. A small part of eastern Santa Cruz belonged to the national park, he told me. The rest was owned by the Nature Conservancy.

Dead ahead, Santa Rosa appeared. Beyond it was much smaller San Miguel, a remote and little-visited place that is mainly home to sea lions and rare northern elephant seals. Like Santa Cruz, Santa Rosa was a patchwork in yellows and light browns, a rugged tapestry of small mountains, dry rolling prairie, and narrow, water-cut arroyos. Rainfall averaged less than ten inches a year, or about half of what fell on the mainland coast. Santa Rosa was home to all sorts of desert plants, such as prickly pear cactus.

There were no good harbors or boat nooks. The most sheltered area was where we were headed, Becher's Bay on the northeast corner. Our pilot delivered a smooth landing on a ribbon of pebbles and sand that paralleled the shore. A mile away was a hardscrabble scattering of utility buildings, barns, and old ranch dwellings. Some were now being used by Park Service personnel, others were still occupied by the former owners. Their deal with the Feds allowed them to continue earning a living by bringing trophy hunters out to hunt the bighorn sheep they had introduced.

A couple of big diesel four-by-four vans stood waiting, one of them for us. It was windy, but also hot. Don Morris wasted no time in getting the beast fired up and the air-conditioning going. Several park rangers were stationed on the island at all times. They kept the vehicles running, the buildings maintained, the roads graded, and they watched over the visitors. Because the only way to get here was by boat or plane, Channel Islands was one of the least-visited national parks in the United States.

Off we drove, past an old ranch house built in the 1870s. Along the shore were pine trees that had been bent horizontal by the wind into amazing shapes, as if from a Japanese painting. Gusts here were often clocked at one hundred miles an hour. Beyond the last barn, we took a sharp turn inland and splashed through a small flowing creek. Morris shifted into four-wheel drive. Soon we were bouncing along an arid track through undulating prairie and trailing a cloud of dust.

Morris filled me in on his career. As a fifth grader growing up in Dallas, he had picked up a book called *Digging in the Southwest.* "I put that book down and said, Gee, I want to be an archaeologist." He entered college at the University of New Mexico and went off on his first archaeology field school after his freshman year. There he heard that the University of Arizona, where Emil Haury headed the department, was a better place to study archaeology. So he transferred, got his degree, and, after military service, came back to Arizona for graduate studies. His very first "real job" was as an archaeologist for the Park Service in Tucson, Arizona. It mainly involved managing and supervising digs by other archaeologists in the parks of the Southwest, but not a lot of hands-on digging himself.

It was a temporary assignment in 1982 that first brought him out to Santa Rosa Island. "I liked it so much that I suggested to the superintendent that he really ought to have an archaeologist on staff here. And that I probably knew where I could get someone to volunteer for this rugged duty." The hint was taken. Morris was transferred to Santa Rosa in 1985 and had been working here ever since. As for the rugged duty, though, he did not actually have to *live* out on the island with his family. There was a ranger's cabin for overnight stays, and he could fly out whenever necessary.

Morris soon discovered that he was walking in the well-trodden footsteps of the late Phil Orr. Based at the Santa Barbara Museum of Natural History, Orr had done all the early archaeological surveys and excavations on the island. Among Morris' first tasks was to read through Orr's many years of field notes.

By the time Orr began poking around, just before World War II, scientists had already found extensive evidence of Ice Age mammals on Santa Rosa. The most notable were the pygmy mammoths, which had gone extinct in the very late Pleistocene. Their bones were plainly visible, eroding out of the canyon walls in many places. The island had also been home to the Chumash people, who still retained a firm and recognized legal interest in what went on here. Morris found himself consulting with them frequently. There were some apparently ancient artifacts as well, but little detailed archaeology had been done. Orr was the pioneer.

On a Sunday in December 1941, just as we had done, Orr flew out to Santa Rosa with a pilot named Bessie Owen in her plane. As he recalled that historic day, they "made a complete aerial coverage of the island, spotting promising fossil deposits, wave-cut terraces, caves, and Indian village sites and returned to the airport on the mainland, only to find that the Japanese had raided Pearl Harbor. This abruptly ended our hopes for immediate work on Santa Rosa Island, which had been so long in the planning, but the delay did have its compensations." Military security meant that no archaeologists or other civilians would be allowed to roam around on these potentially strategic islands. But he *was* permitted to spend many hours flying over them as an observer in military aircraft.

After the war, Orr got down to systematic archaeology on Santa Rosa. He flew out almost daily in the spring of 1947 to conduct aerial surveys, and then covered the more promising areas on foot. It was tough going. There was only one narrow, twisting, washed-out trail along the ridge of the mountains, and all travel was by foot, horseback, or boat. Later, Orr brought a Jeep out to Santa Rosa. Morris joked about how cushy he's had it by comparison. Orr had to maintain the Jeep himself, change the oil, clean the carburetor, adjust the points, replace the tires. "If this truck breaks down," said Morris, "I'll just get another," and the servicing is all done by park staff.

Late that first summer, Orr set up a camp and began excavating and collecting pygmy mammoth bones. He found what he called "a repeating pattern" of sites that had both burned mammoth bones and areas with charcoal in them. Some bones seemed to have been butchered. He also found abalone shells fairly deep in deposits that were well away from the shore, which seemed to him to be evidence of human activity. How else had the shells been carried so far inland? Finally, there were some chipped stone tools. During the 1950s, he had some of the charcoal dated to between 12,500 and 29,700 years of age. He concluded that people had been on the island during, and perhaps throughout, the last glaciation, and had likely coexisted with the mammoths. This was a highly controversial notion.

For two decades, Orr spent anywhere from one to four months each year on the island, excavating mammoths and traces of Indian life and focusing mainly on the island's northwest coast, where Morris and I were headed. "This area," he wrote, "constitutes one of the best natural laboratories in the world for the study of the late Pleistocene and Recent [Holocene] periods, with its many wave-cut terraces, fossil dwarf mammoths, submarine geomorphology, and a continuum of man's habitation during the past 10,000 years or more, represented by twenty-two village sites in this one area of about 200 known on the Island."

Then came the find that really put Santa Rosa on the archaeological map. Every year since the late 1940s Orr had been prospecting along the steep bluffs of Arlington Canyon for mammoth bones and anything else that might be interesting. One day in 1959, he was working his way across and down one of the canyon walls with three assistants when he noticed a bone protruding from a cut bank. It was more than thirty feet below the grassy and almost level surface at the top of the canyon, and less than half a mile from the sea. He recognized it immediately as a human femur, or thigh bone.

A little probing showed Orr that there was more than just the one femur. Eventually, it turned out that the spot held a second femur, plus a humerus (the longest arm bone), and a fourth bone that he could not identify. Orr cut away at the soil in the bank around the bones, to see as much as he could. Then, because the rainy season had begun, he and his team covered the bones with metal foil, plaster of paris, and burlap, and built a weatherproof casing to protect them.

He did not want anything to be moved, removed, or damaged until he could get some of America's top scientists to come out to Santa Rosa. He needed expert witnesses who could attest that the bones had been found *in situ*. It took all winter to organize, but the next year, 1960, a major field conference was held on the island.

Morris and I had now careened and shimmied about twenty miles across the parched landscape to reach the island's northwestern shore. Arlington Canyon was a windswept, treeless, deeply eroded cut about a quarter mile wide and several hundred feet deep. Morris parked on scrubby grass at the top of the canyon, where there was a hazy view of the sea. A light breeze whispered in off the water. We donned our daypacks, and he led me to a spot overlooking the steep dropoff and the creek far below. He pointed down along the canyon slope to the excavation area. I was amazed that Orr had spotted the bones in the first place. The vista was a vast and dazzling patchwork of brooding cliff faces, huge boulders, bare expanses of red earth, meandering gullies, and low bushy foliage along the creek. "My understanding," said Morris, "is that he was just walking up the canyon, undertaking a general archaeological survey. When you turn an archaeologist loose in his field area, he almost always has an eye out for something fresh that he hasn't seen before."

Just then, something caught Morris' attention. He pointed across the canyon to a tiny spot that stood out mainly because of its color. It looked to him like a possible mammoth bone lying in a deep gully. "From up here you can spot quite a few things," he mused. "If we were out here looking for mammoth bone, I would go over and investigate it. It's the right color and about the right shape," although without getting much closer, he could not be certain. "Not many people can just say, oh it must be a mammoth." The exception was tusks. "Tusks can be identified. Tusks stand out at a distance." The canyon, though, was rich in mammoth remains. He himself had found around ten specimens in this immediate area.

Morris went on with the tale of Phil Orr and his dramatic Santa Rosa find. "He realized pretty clearly that he had something exceptional, because he didn't just pull them out of the bank. He exposed them, looked at the geology, and realized that he was down quite far below today's soil surface." This meant that the bones could well be quite old. "He left them in place and arranged the Arlington field conference. And he got most—nearly all—of the big guns in Early Man archaeology out here to the island." The most prominent were James Griffin, Emil Haury, and Alex Krieger. They also looked closely at his other, related evidence. "This included the fire hearths and what Orr interpreted as butchered pygmy mammoths."

Emil Haury, one of the "big guns," was at the time the head of Don Morris' department in Arizona. Looking back, Morris marveled at how a site like this could be passed down from one generation of archaeologists to the next. Haury came out to Santa Rosa in 1960, and now, even in 2000, the site was still being measured and analyzed. Within the next few weeks, in fact, further study was going to be done of the embankment where the bones had been embedded. "Work on this thing has been going on my whole career," Morris laughed, shaking his head. "And here I am. I'm 115 days away from retirement. And my last field project is going to be out here on the Arlington site. So, we've spent nearly half a century screwing around with this thing. That's what it amounts to. And we still don't know. . . ."

During the winter before the field conference, Orr managed to have a radiocarbon date done on a sample of soil that had been in direct contact with the bones. There were particles of charcoal in it, but not a whole lot. The dating came out to 10,400 years, but because the organic sample was so small, the error margin was huge (plus or minus 2000 years) and therefore inconclusive. This meant that everything depended on the conference. It was Orr's big chance, his only chance, really, to impress the mainstream of American archaeology, geology, and paleontology.

They came, they saw, but they were not conquered. Morris had read Orr's notes from the time of the conference. "And what he wrote about it—what Orr was hoping for—was that these people would come out to this island, look at what he'd found and react in a manner similar to what happened at the Folsom Conference in the 1920s. When people looked at the material in the field and realized, *Hey, we have people going back thousands of years, hunting extinct bison.*"

"He hoped they would say, 'By golly, Orr, you're right, we've got Pleistocene Man on Santa Rosa Island.' Well, people came out, and they liked what they saw. They spent a lot of time here." But they also saw problems. One was the real possibility that the block of soil in which the bones were encased had actually slumped down, that instead of belonging thirty feet or more below the surface at the top of the canyon, it should have been only five or six feet deep. This would make it just another burial, of which there were thousands on the island. They also looked at the pygmy mammoth bones and the so-called "fire hearths" and they were not convinced. Burned bones could have been the result of naturally caused fires.

"Judging by the letters he wrote, Orr was really disappointed at the outcome of the conference. Because I think he was hoping that they would lift him up on their shoulders and say, 'Yes, you've made a huge breakthrough here in our understanding of American prehistory.' Instead, they just didn't buy it. And people since then have not bought into the notion of Pleistocene hunters pursuing the pygmy mammoths out on Santa Rosa, 20,000 or 30,000 years ago."

There was still the possibility of getting better radiocarbon dates, and the next year another attempt was made. Two independent geochronologists from Columbia University collected new samples about a foot away from the bones and got a date of 10,000 years plus or minus 200. In some ways, this confirmed Orr's hunches. The block of soil had *not* slumped down to that depth. It really *was* old, not relatively modern. But ultimately the inconclusiveness of the conference and its follow-up knocked the wind out of his sails, and he never even completed the dig or published the results in a complete professional report.

"Orr wrote some very bitter things in his notes. And I think the whole thing was a real disappointment to him. And I'm guessing here," said Morris. "This is all speculation on my part. I'm trying to read his mind, which is kind of baloney. But if you believe that you have people out on the island 30,000 years ago, then finding a 10,000-year-old skeleton is nice, but it's no big deal. And that, I think, is why, having gotten his day at the conference, Orr put the bones in the back of the museum and didn't do anything with them. Which I'd say was fortunate for science in the long run.

"After that conference, Orr really didn't do much of any fieldwork out here on the island. As I interpret it, that was the death knell to his hopes for acceptance of his idea about pygmy mammoth hunters." Aside from a couple of short journal articles that he wrote soon after finding the bones, Orr's only major publication was a 1968 book, an overview of the prehistory of Santa Rosa Island. "And his data is just not presented in enough detail to allow you to confirm or deny his ideas. That was his Achilles' heel. He didn't publish in enough detail." He never even published a proper map of the Arlington site, or the complete photos and descriptions that would allow future archaeologists to decide whether his interpretation was right or not.

We scrambled down the hillside, and Morris showed me where the bones had come from. In the next month, he told me, a couple of geologists and geochronologists would be coming out to take more soil samples from the area close to the bones. But they would not be able to pinpoint or study that spot precisely. After the field conference, Orr had had a bulldozer cut away at the bank to carve a road down to the site, and in doing so he moved back the entire face of the canyon. "The bones were about where my belt buckle is right now," said Morris, standing five or six feet from the vertical wall of dirt and rock.

As for the bones themselves, Orr jacketed them in plaster, but they were still *in situ*, still well encased in their block of surrounding soil. He pulled the entire thing, intact, out of the embankment and took it back to the Santa Barbara Museum. The block, said Morris, was about three feet long, eighteen inches high and about eighteen inches to two feet deep. And the bones sat in the museum, neglected, until Morris arrived on the scene.

"When I came out here to Santa Rosa, at first still only on temporary assignment, I ran into the ranch foreman. 'Oh, you're an archaeologist.' he said, 'I've got something I'd like to show you.' " It was a deeply buried set of human bones that were eroding out of a sea cliff at another spot on the island. "This was when AMS dating was just being established." Morris had minute samples of the bone dated at the lab at the University of Arizona, and they proved to be 8600 to 8800 years old. "That got me thinking about Arlington Canyon. I went to John Johnson," who had recently become curator of anthropology at the Santa Barbara Museum, "and said, let's find Arlington Man and do AMS dating on it. We can probably get dates on the bones without destroying the whole thing."

Morris was being modest about his own role. (He is listed as coauthor of the professional paper that eventually resulted from the redating of the Arlington bones.) "It was kind of a no-brainer to think, *What about Arlington?*" he said. It looked like it might be very old, but the dating of around 10,000 years had been criticized, and rightly so, because the dates were from the charcoal around the bones. "Whereas the most conclusive date would clearly be on the bone itself." Morris and Johnson found the plaster block, clearly labeled, tucked away in a dingy corner of the museum's subbasement.

"I always say, the reason I'm coauthor of the paper with John is that I used my torch to get the tarantulas off his back."

Fortunately, Morris had some Park Service money available to contribute to the redating project, which ran to about $6000 in all. They sent bone samples to the same lab in Tucson that Morris had used earlier, at the University of Arizona. But that lab decided there was not enough collagen in the samples, the bone protein needed for accurate dating. The lab could supply a date, but the error margins would be large. "We said, *We've got enough crummy dates already.* So we waited a little bit." This was only the first in a long series of detours and false hopes. A number of other labs had a go at the samples, trying a variety of techniques. But even after six years, none had succeeded in purifying the collagen enough to achieve an accurate date. Meanwhile, though, further excavation of the block of soil was done to expose the femurs. Anthropologist Phillip Walker of the University of California in Santa Barbara took measurements that indicated the bones were probably from a female. So the specimen was now Arlington Woman, not Man.

"Then Tom Stafford came into the picture." Stafford was among the country's most respected geochronologists. It was Stafford who also dated the human bones from On Your Knees Cave, Alaska. The method Stafford used to isolate pure enough samples of collagen had so many steps to it, each of which had to be executed with such precision, that it was the closest thing in archaeology to rocket science. And then came the nuclear physics. The isolated and purified sample had to be bombarded in a particle accelerator.

In the end, though, Stafford came up with a date he could publish with confidence: 10,960 plus or minus eighty years. In other words, essentially 11,000. And he had a nice way of checking on his result. Also encased in the block of soil were the bones of a now extinct species of deer mouse, including a well-preserved mandible. Because the mouse bones were considerably better preserved than the human ones, they gave him a purer sample of bone collagen. The date from the deer mouse was 11,490 plus or minus seventy years. Throughout the long, step-by-step dating procedure, Stafford found that the higher the chemical purification of bone collagen he could achieve, the *older* the dates turned out to be. His overall conclusion, therefore, was that the true age of Arlington Woman is probably somewhere between 10,960 and 11,490. In other words, 11,000-plus years.

Stafford's date may not be the end of the story. The bones have not been destroyed. They are still preserved in their original matrix, which means that if, someday, a further advance is made in dating technology, the entire issue can be revisited. "This is a textbook example of why you need museums," said Don Morris, "and why you need to preserve things and keep them. Because you just don't know what you're going to use down the road. New techniques, new perspectives, you name it. It's worth hanging on to this stuff." And the fact that Orr had left the bones undisturbed, *in situ* in its original block of soil, was crucial. "If he had cleaned them and removed them, they would no longer have been in association with the mouse bones."

Morris had spent years thinking about the fate of Arlington Woman. Who was she? Why had she died here? What was she doing here? The spot was hundreds of feet above today's sea level, and would have been a lot higher at the time of much lower world sea level when she lived here. "Too bad it's so hazy today," he said, pointing out across the water. San Miguel Island, which we had glimpsed from the plane, lay out in that direction. His personal hunch was that Arlington Woman had been out hunting or foraging from a major cave and rock shelter called Daisy Cave on San Miguel, which would have been part of the same giant island at that time. "It's only about four miles away."

In the mid-1980s, Morris was part of a team that began excavating inside the cave. They found early Holocene artifacts dating back as far as 8500 years. Then Jon Erlandson of the University of Oregon did further work there and found evidence that the lowest levels in Daisy Cave may have been occupied as early as 10,400 years ago. But the cave continued to be used by the Chumash right up into historic times. Morris called it a "relict" sea cave that stood well above sea level both in modern times and during last glaciation. It might have been carved out of the cliff by wave action a million or more years ago. But since it is located on a very steep shoreline, no matter where the sea level was, it would still be very close to the sea. "That's one reason it's got this tremendous span of occupation."

But why suspect a link to Arlington Woman? The key word, and the attraction as he described it, was "cave." It was and still is, an excellent and well-located source of shelter. "The first time I saw it, I was walking along and it was blowing about thirty knots or so, and I thought, *Hey, I'm going to remember this spot, because if I'm ever out here, say, and I can't get back to the cabin, this is a good place to bivouac. I can at least get out of the wind.* I think that's been its attraction throughout several millennia." It was occupied into historic times, "but the lower layers are basically contemporaneous with the Arlington find here."

We made our way to the truck. It was time to head back across the island to catch our return flight to the mainland. On the way, Morris offered his gut hunches of what the archaeology on Santa Rosa indicated.

For starters, he was highly skeptical that the people here were ever mammoth hunters. On the other hand, it was entirely possible that people and mammoths were here during the same relatively short time period. There's a big difference, he said, between simply coexisting and arguing, as Orr did, that "you had a culture out here that lasted for millennia on the islands with pygmy mammoths as prey. When you think about that in terms of what we know about island biogeography, and what we know about human culture, that's quite a claim, because this is a small island." In biological terms, it would have been a relatively small island even when it was ten times larger.

The elephants were at the top of the food chain, he went on. There would have been relatively few of them, each consuming vast amounts of vegetation over a very restricted territory. He thought this made for "an inherently unstable situation." "You might have a period where the people hunted the mammoths to extinction. That's what happens on islands. You have moas or dodos or something, and people get there, and pretty soon they've used up all the dodos and they turn to something else." But "there was just no way that people could have been hunting mammoths for thousands of years," as Orr seemed to believe. "What I think is that the first people came out here, and they may have seen the last of the pygmy mammoths walking around here." That would, of course, imply very early, and certainly pre-Clovis, coastal migration. Recent results using AMS radiocarbon dating showed the "youngest" pygmy mammoths to be around 12,840 years old. If people wiped out the last of the mammoths, the time frame would be just about right for a coastal migration that took people to Monte Verde by 12,500 years ago.

But Don Morris was not sure that coastal people would have hunted the mammoths at all. "If they were boat people, used to gathering food from tidepools, or doing abalone and inshore fishing, with netting techniques, and using bone gorges to get their meals, they probably looked at these big animals and decided not to mess with them." How, he wondered, would they tackle such great beasts, galumphing across the landscape?

"These pygmy mammoths were respectable-size animals. They stood about six feet high at the shoulders and weighed about a ton." People would not go after them with seal fishing harpoons. "They might have said, 'Oh, I think I'll leave them alone.' But they were probably dealing with a population that was on its last legs anyhow," due to rapid late Ice Age environmental changes. "It would be interesting to know if the first people here at least saw the mammoths."

I thought about Arlington Woman on the way back to the airstrip. The redating of her bones was a major breakthrough for the coastal migration scenario. Until then, all of the remote offshore artifacts (in Alaska, in the Queen Charlottes, and Jon Erlandson's finds at Daisy Cave on San Miguel Island) were significantly more recent than Clovis. In fact, they were as much as a thousand years more recent than the very *oldest* Clovis. In theory, at least, that meant that inland Clovis people could have made their way to the coast. While living there, say for a few hundred years, they could have developed boat-building technologies and maritime hunting adaptations and made their way out to the islands.

It did not seem very likely. After all, why would they give up their inland big-game hunting practices, migrate across the mountains to a very different climate zone, and then head out to live, hunt, or forage on offshore islands? What made it even less likely was the fact that the Folsom culture had succeeded Clovis. Big game hunting had not died out. It had continued for at least another thousand years across the American Southwest. Still, the hypothetical scenario in which Clovis was a forerunner of the offshore coastal peoples could not be ruled out.

But now, here was Arlington Woman. She was nicely contemporaneous with Clovis, which lasted from around 11, 200 to 10, 800 years ago. And she was out here on an offshore island reachable only with well-developed boating technology. There was no longer any reason at all to assume that the land-based Clovis technology preceded the maritime one. A major conceptual barrier had been swept away.

Don Morris was also impressed with the implications of Arlington Woman. The idea of very early coastal migration, he thought, "looks more and more appealing, the more we substantiate earlier human presence down here in California. We're on an island. We're very confident that it's always been an island. To get to this island you have to cross the Santa Barbara Channel. Eighteen thousand years ago it was a lot narrower, but I'm not sure that that meant it was necessarily easier to cross, because the narrower channel implies swifter currents. In a way, it was like a big, broad, nasty river. It was still five miles wide with a strong current. So they had to have some means of water transport. And you look at the Australian situation, where you had people making island-hopping trips 40,000 years ago. Even though we haven't found the direct archaeological evidence, I think people here had boats in their tool kit at the end of the Pleistocene."

To him, the coastal migration theory just made so much sense. "I haven't studied it deeply. I'm a local boy here, and I've got my nose in the dirt out here in these islands. But just think about the pathway that exists along this coast. The kelp beds, the tidepools." He stared out at the water as we sat, waiting for the plane to arrive, and tried to visualize those early coastal people. He imagined them up in Alaska, "looking at the otters, looking at the abalone, looking at the tide-pools," and learning how to hunt and gather food along that shore. There would have been no barriers at all to movement down into new coastal territories. "All they had to do was to keep following these easily harvested resources southward. And the biggest change they would experience," he laughed, "was that the weather would get more and more comfortable."

CHAPTER SEVENTEEN

Emerging Consensus

I LEFT Santa Rosa Island more convinced than ever that coastal migration was the most credible explanation for the first peopling of the Americas, but I felt compelled to put some more meat on the bones of what was still very much the mere skeleton of a theory. I also had to extend my vision southward. Until my California visit, I had focused on the northern Pacific rim, where the greatest challenge was to explain how people had managed to skirt the ice at a time when much of the mainland shore was still locked under thick glaciers. Island hopping from one offshore refuge to the next seemed to be the answer.

South of Washington's Olympic Peninsula, though, there had been no great ice sheets flowing down to the coast. Moreover, migrating people would have found no significant islands for more than a thousand miles, until they reached the large Ice Age island of Santarosae off today's Santa Barbara. So I grappled with the question of why people would have remained on the shore. Why would they continue moving southward along the coast rather than spreading eastward and inland?

One clue came from a new book on Haida legends and the world they depicted, *A Story as Sharp as a Knife*. Author Robert Bringhurst wrote that Haida storytellers "speak of three distinct realms—forest, sea and sky—each with its native population. None of these, however, is the human realm." Each of these worlds is inhabited, instead, by a bizarre array of animals and mythical creatures, many of them lurking in dense forest or beneath the surface of the ocean. "Humans are only at home on the *xhaaydla*, the boundary or intertidal zone, at the conjunction of all three. A few strokes of the paddle or a few steps into the bush are enough to leave the human world behind." This boundary zone, the richly endowed coastal strip, is extremely narrow. Along much of the coast from Washington to California, for example, high cliffs and steep mountains plunge right down to the sea. Farther north, the modern Haida, even after living for thousands of years on the edge of the forest, still seemed to find any area beyond the beach strip somewhat alien. All the houses in their villages were situated within a hundred yards of the sea, and in many cases, much closer. What would ancient coastal peoples have thought about the dark woods that fringed the beaches where they had learned to live for generation after generation? Even where the forces of nature—erosion, flooding, and wind—created more open areas, such as along broad river valleys, what would motivate people to head inland?

Among their minor worries would be much higher concentrations of biting insects than on the seashore. More significantly, they would have to face large Ice Age beasts, some of them incredibly fierce. The giant short-faced bear was a long-legged killing machine that could run like the wind. There were also deadly saber-toothed cats and huge dire wolves with teeth several inches long. Grizzly bears and North American lions were other animals that maritime people armed with bone-tipped harpoons and wooden spears would be reluctant to confront. The less aggressive large animals would have been formidable prey as well: mammoth, mastodon, bison, horse, and western camel, an extinct beast that was different enough from modern ones that scientists assign it to a separate genus. To hunt these animals, maritime people would have to develop entirely new and different weapons, such as large spear points. This would require finding new sources of specialized stone. And as people continued their trek inland, they would have to keep locating quarries of such stone in each new region they colonized.

The immediate coast, by comparison, would have been relatively safe and familiar, and full of easily harvested food. Shellfish could be dug in vast quantities from any beach or, in the case of mussels, picked right off the rocks at low tide. Seaweeds fringed the shore in such densities that kelp beds have been called "forests of the sea." Fish could be caught in nets from the beach, or in traps at the mouths of rivers and streams. Eggs could be collected from huge bird colonies. There were also large, but nonthreatening, animals, such as seals, sea lions, and the huge Steller's sea cow, a vulnerable browser related to the dugong and manatee that was only hunted to extinction about 200 years ago. An occasional dead whale, drifting ashore, would provide a windfall of food, oil, and useful whalebone.

The coast also offered ever-replenished supplies of driftwood, from trees washed down continental rivers. Much of it would already be broken by the action of the sea into manageable pieces, washed up onto beaches by high tides and storms, and left there to dry in the sun. Driftwood would have provided fuel and material for shelters even on the treeless Ice Age shores of the northernmost Pacific. But coastal resources were abundant, not unlimited. And in such a narrow belt of habitation, they would eventually become depleted, especially as populations expanded. However, for the first people moving southward along the West Coast of America, there would always be an untouched coastal fringe just a bit farther along the way. Hence the motivation for continuing, and possibly quite rapid, migration. Ahead, acting as the carrot, would be more bountiful virgin beaches full of shellfish and lined with driftwood. Behind the avant garde of migrants, serving as the stick, would be a growing population of hungry people competing for ever-scarcer resources.

At every stage in the process, at least until the entire coast was colonized, it would be easier and safer for people to cling to their accustomed coastal adaptation rather than to venture inland and challenge the forests, mountains, and deserts. In this way, people could have populated the narrow coastal strip right down to southern Chile in something like one or two thousand years, with hardly any serious push into the interior of either continent.

Rapid coastal migration could account for the 12,500- to 13,000-year-old sites like Monte Verde and Taima-Taima, and also for the apparent lack of human remains in the Americas older than 11,500 years. The drastic late Ice Age rise in sea level provided a simple and entirely credible explanation for the limited offshore archaeological evidence that had turned up so far. Since nearly all of the best places for ancient coastal people to gather food during the waning millennia of the Ice Age were now under hundreds of feet of water, it was in no way surprising that archaeologists had found few traces of such a population. In fact, considering how little time and resources had gone into the coastal search, what had been found was highly impressive. In the late 1990s a consensus began to emerge favoring coastal migration as the most likely explanation to replace the toppled Clovis First paradigm.

Well-dated hard evidence from coastal archaeology and its allied sciences (geology, paleontology, and biology) was convincing in a way that no amount of theorizing based on linguistics or genetics could possibly be. And coastal occupation was not limited to the Pacific shores of North America. If coastal migration were to account for the first settlement of South America as well, archaeologists should find indications of very late Pleistocene coastal habitation along the long stretch between California and southern Chile. Rising sea level would have inundated the late shoreline there, just as it had in North America, obliterating most potential sites. But, as on Prince of Wales Island and Santa Rosa, people might be expected to leave at least some scattered traces inland and above sea level. Yet some experts on South American archaeology doubted that people had begun to exploit marine resources there until the Holocene, that is, after 10,000 years ago.

As if on cue, evidence of earlier maritime adaptation came to light in 1998 from two sites in Peru, Quebrada Tacahuay and Quebrada Jaguay. The latter was occupied from 10,000 to 11,100 years ago, and the former for more than two centuries during the middle of that time span. Quebrada Tacahuay would have been about half a mile from the sea when sea level was low, and it revealed fire hearths with fish, bird bones, and other evidence suggesting net fishing and the butchering of sea birds. Quebrada Jaguay would have been four to five miles inland along a stream. It contained clamshells and fish bones carried up from the coast, along with stone tools, including obsidian brought down from the Andean highlands. These sites added welcome confirmation to the coastal scenario.

Another important development came from a cave on the Queen Charlotte Islands. Paleontologists had long supposed that the Charlottes, along with the outermost islands in southeastern Alaska, were glacial refuges for bears during the peak of glaciation. But they had no proof. The record of mammal bones found by Tim Heaton and Fred Grady on Prince of Wales Island left a significant gap. No bear bones there dated to the time from 12,300 years ago to about 35,000. There were some seal bones dating to around 14,000 years and from around 17,000 to 21,000, but these could have been scavenged down at the sea and brought up to the cave by foxes.

In 2000, foresters began surveying a wooded area in a remote part of the Queen Charlottes in advance of possible logging. Two of them had discovered a large limestone cave there years earlier, and a consulting firm led by caver Alan Griffiths was employed to examine it more closely. It turned out to be spectacular, with a main entrance almost 200 feet wide and a waterfall cascading down into its maw. The main passage was about 500 yards long, with ceilings thirty to sixty feet high and a stream flowing right through it. Most importantly, it contained the bones of at least six different black bears. When the bones were dated, the oldest proved to be 14,500 years old. This pushed the date of known bear habitation on the offshore islands more than 2000 years farther back into the glaciation.

In 2001, Griffiths returned with Daryl Fedje for several days to map the cave further. It proved to be larger and more complex than initially thought, with other entrances and a number of side passages worth exploring. Fedje was understandably excited by what else might be hidden in the deposits of sand and gravel, and announced plans for a much more systematic study of the cave. "With any luck," he said, "there'll be some species other than bear."

Just how the earliest coastal inhabitants fit biologically into the larger human family, and where they may have originated, were other questions that remained unanswered at the cusp of the new millennium. The late Pleistocene coastal artifacts and human remains found so far are pitifully few, and they provide more questions than answers. The studies of mitochondrial DNA remain highly contradictory. Some geneticists think the evidence points to three or four separate migrations occurring over a long span of time and from widely separate parts of Asia; others see only a single but very early migration. The long chronology endorsed by so many geneticists (and a few linguists as well) is at odds with the most solid evidence from archaeology, especially the absence in the New World of human remains older than 11,500 years, and of well-dated artifacts older than 12,500 to 13,000.

Physical anthropologists, using evidence mainly from measurements of ancient human skulls, also weighed heavily into the first peopling debate in the late 1990s, especially in the wake of the Kennewick Man discovery. Most skulls older than about 8000 years found in North America do not *look* much like the skulls of either modern Indians or Asians. The few ancient heads that are complete enough to give an accurate picture are long, with narrow brain cases and narrow faces and noses; some have receding cheekbones and high chins. These are just the characteristics that forensic anthropologist James Chatters labeled "Caucasoid" when he first examined Kennewick Man. By contrast, American Indians and modern Asians tend to have rounder heads and broader faces.

Many physical anthropologists now believe there were at least two (and possibly more) influxes of quite different people into the Americas, including an early wave that brought the long-headed people and a later one that introduced people with more Mongoloid features. Others attribute the differences between ancient American skulls and more modern ones simply to "genetic drift," or normal evolution occurring in a population over the course of 10,000 or more years. A few anthropologists, notably the Brazilian Walter Neves, argue for an influx of people to South America who resembled modern Africans or Australian Aborigines.

But where could the supposedly "Caucasoid" early North Americans have come from? One proposed answer fits quite well into the coastal migration scenario.

C. Loring Brace, curator of the University of Michigan's Museum of Anthropology, has closely studied the ancient Jomon people of Japan and their modern descendants, the Ainu minority who still live on the northern island of Hokkaido. When he first read the *New York Times* article announcing the Kennewick find, he stared at the picture of the skull and recognized it in an instant. *I know who that is*, he said to himself. *He's a Jomon.* Brace was appalled at the idea that the Kennewick Man skeleton might be buried by Washington State Indians and lost to science. He wants the chance to get his calipers onto that skull and find out whether his theory of its Jomon connection is correct. So he joined the lawsuit against the U.S. government as one of the eight prominent plaintiffs.

The Jomon share many physical characteristics with Caucasians, but Brace says that they are a separate genetic stock. They date back at least 15,000 years in Japan. Jomon archaeological sites have included fish hooks, harpoons, and other signs of maritime adaptation that date back at least 9500 years. Pre-Jomon people in Japan had enough maritime skill to bring obsidian to Honshu from offshore Kozushima Island over 30,000 years ago. Assuming that the Jomon, too, had adequate watercraft by 15,000 years ago, an offshoot of their population would have been well situated to begin an early migration around the North Pacific rim. Physical anthropologist D. Gentry Steele, of Texas A & M University (also a plaintiff in the Kennewick lawsuit) suspects that "the first colonizers will prove to be populations from southern or central Asia, such as the Jomon of Japan, although the evidence is as equivocal for this as for other views."

There is also a well-documented inland archaeological site with a wealth of advanced stone tools at Ushki Lake on the Kamchatka peninsula, well to the north of Japan. Those tools date to over 14,000 years ago, and possibly to 15,000, which makes people from Kamchatka likely candidates as early coastal migrants as well.

Of the possible ways to account for the first peopling of the Americas, coastal migration late in the Ice Age is by far the most compelling, and the most widely endorsed today by paleoanthropologists. Other ideas and explanatory models continue to vie for attention. But all have major weaknesses.

Some supporters of a long chronology for the New World have proposed that people might have arrived by a Bering Strait overland route well *before* the peak of the last glaciation, 30,000 or more years ago. At that time there would have been no coalescing ice sheets to obstruct migration. Such an early overland migration might account for the claimed 30,000- to 50,000-year-old artifacts at several sites, mainly in South America. And this scenario could account for the linguistic and genetic diversity found in the New World, which, some specialists in those fields have argued, could never have emerged in "only" 12,000 to 15,000 years.

The problem, though, is that there is no independent way to verify that languages in the Americas, or genetic lineages for that matter, have branched off and increased in distinctiveness at a similar rate to such evolution elsewhere. As Jim Dixon said, these are "noble" attempts at theorizing, but they can never be convincing in the same way as solid archaeology. As for the touted very early archaeological sites, such as Pedra Furada in Brazil, none of them has satisfied the highest standards for proof of human presence and dating. Finally, and perhaps most indicative of all, no well-dated human remains significantly older than 11,000 years have been found on either continent. If people came 30,000 or more years ago by an overland route, at a time when sea level was nearly the same as today, their remains should be scattered widely across the interior of both continents.

Yet if they were really here between 15,000 and 50,000 years ago, why has nobody found *any* solid evidence of them? There should, in fact, be plenty of it. In Europe, for example, there is a vast body of undisputed Ice Age archaeology from that time period, including unmistakable stone tools, mammoth kill sites, cave paintings, and lots of Neanderthal and Cro-Magnon fossils. Particularly the absence of verified pre-Clovis human remains in the Americas, even after so many years of searching, underscores this conspicuous void in the archaeological record.

Coastal migration with a relatively short chronology suffers from no such weakness. Assuming that maritime people reached the New World 13,000 to 15,000 years ago, it is quite understandable that they might have remained mainly on the coast for the first few millennia and not appeared in the interior until around the time of Clovis. Thus the absence of artifacts and human remains in that time period is easily explained, as Knut Fladmark argued from the start.

Another long chronology scenario has people reaching South America well *before* they colonized North America. The most likely way this could have happened is by means of long transoceanic voyages from southeast Asia or the South Pacific. (A few South American anthropologists also argue for possible direct transatlantic voyages from Africa.) This might explain how Monte Verde, Pedra Furada, and perhaps other disputed South American sites could be much older than known sites in North America. Even Jim Dixon insists that we keep an open mind about such migrations. Tom Dillehay, too, seems quite sympathetic to this idea.

But the winds and currents in the southern and equatorial Pacific are predominantly from east to west, and are therefore highly unfavorable to such voyages. This makes a very long accidental drift voyage from the western Pacific to South America unlikely. (As Thor Heyerdahl demonstrated with his Kon Tiki experiment, it is much simpler to sail, or even drift, westward from South America to Polynesia, than the other way around.) When Polynesia was actually colonized, only in the last 2000 years, it was done step-by-step from west to east, one island group after another. The Polynesians did it systematically by sailing upwind and against the main currents. This was a sound strategy for people who were intentionally seeking new islands to settle. They were not just boatloads of people out fishing, who happened to be swept away by wind or currents. They brought domesticated plants and animals with them. They sailed off in large enough numbers to establish new colonies. If on any given voyage they failed to find new islands in the trackless wastes of the Pacific, they could simply turn around and sail downwind. This would bring them quickly and safely back to their known home islands.

* * *

What also casts doubt on colonization by means of long direct voyages to South America is the sheer number of people of childbearing age required to establish a stable new population on an isolated island or empty continent. It is highly unlikely that this could be done by the proverbial lone pregnant woman surviving on a raft, or even by an Ice Age Adam and Eve. If a very small number of people reach a new territory, during the first dozen or so generations there is a high likelihood that a short string of "bad luck" will occur and lead to the population's extinction. (The unlucky events can be as simple as bearing mainly all male or all female children in a particular generation, or having a higher than average death rate for an extended period of time.)

The late Australian mathematician Norma McArthur and two colleagues studied this problem statistically to understand how many couples of childbearing age would have been required to populate a remote island in Polynesia. Using data from studies of aboriginal Polynesians, they plugged in such factors as life expectancy, fertility rates, age of puberty, and the survival rates of babies and young children. They introduced randomness and ran computer simulations hundreds of times to model the process of population expansion and its alternative—crash and extinction.

What they discovered to their surprise was that tiny populations could struggle along for hundreds of years and still eventually crash (due to an unfortunate chain of events) and go extinct. Generally, they were not "safe" and established until they had expanded to at least thirty to fifty people. As for the initial founding group, if there were only three men and three women of childbearing age, they would most likely fail to become established and go extinct within a few generations. Starting with five men and five women, the odds became better than even that they would survive as a population, but only if the women were in their very youngest childbearing years (with many years left to procreate) when they arrived. With seven men and seven women, the odds again improved slightly, but, unless the women were very young, there was still a 35 to 40 percent chance of eventual extinction.

In other words, a minimum of ten to fourteen individuals would have to reach South America at once to have a reasonable chance of establishing a permanent colony. And this is only if they were all of ideal age and evenly divided by sex. If not, a larger initial number would be required. Furthermore, this study did not take into account the dangers of inbreeding and lack of genetic diversity within such a tiny group. (Looked at another way, when the number of individuals of any mammal species on an isolated island—or in a similarly isolated terrestrial habitat, such as a mountaintop ecosystem—falls below several dozen, they usually go extinct.)

How likely is it, then, that ten or more people, including at least as many young women as men, would have survived a long drift voyage 30,000 or 40,000 years ago? First, they would all have to be out on the sea together on an extremely well-provisioned canoe or raft. (By definition, a long accidental drift excludes a coordinated flotilla of smaller watercraft.) In the course of a voyage of thousands of miles across the scorching tropics, they would have no access to fresh water, except for any rain they could catch. If any of them failed to survive, then the starting number would have to be even larger. This implies quite a substantial craft.

For comparison, Thor Heyerdahl's Kon Tiki sailing raft, modeled on the kind used along the west coast of South America when the Spanish arrived, was forty feet long and barely able to carry food, water, and six men. Were people in the Solomon Islands building and outfitting watercraft that size or larger more than 30,000 years ago? It seems unlikely. They would not have needed anything like that to travel among their local islands. Nor is it probable that they had invented sails yet. More likely, they paddled small, lightweight rafts or dugout canoes.

Conceivably, a very ancient migration to South America could have involved island hopping, and thus a series of much shorter unplanned voyages over a long period of time. But there is a problem here as well. No evidence of such voyages has ever been found on the intervening islands of Polynesia. Possibly the artifacts have simply not yet been found, but there is an additional difficulty. When people finally *did* reach Polynesia for certain, about 2000 years ago, their sudden presence left unmistakable traces. They not only brought along distinctive pottery and other artifacts, they rapidly wiped out many species of native birds and animals, while introducing other new plants and animals. It is hard to imagine how island-hopping voyages tens of thousands of years earlier could have taken place without leaving similar traces. And there is additional evidence that the first spread of people across the southern Pacific was relatively recent. Geneticists have studied the mitochondrial DNA of native Pacific rats, for example, and have shown that the rats reached remote Polynesian islands at roughly the same time the Polynesian people did, not many thousands of years earlier.

For all of these reasons, the small corps of mainly Australian and New Zealand anthropologists who have studied the techniques, patterns, and timing of early Pacific voyaging reject the notion of very ancient trans-Pacific migrations.

A different type of long, colonizing voyage has been proposed by Dennis Stanford of the Smithsonian and his colleague Bruce Bradley. They think that pre-Clovis people with a stone tool technology that is somewhat akin to Clovis could have come by boat from southwestern Europe between 16,500 and 18,000 years ago. The hypothesis is intended particularly to account for similarities they see between Clovis spear points and so-called "Solutrean" ones from Spain and southern France that predate Clovis by 5000 or more years. It could also account for artifacts found beneath the Clovis level at several archaeological sites in the eastern United States, and possibly also the rare and anomalous "X" mitochondrial DNA lineage, which is found among some 3 percent of Native Americans.

Stanford and Bradley envisage these Ice Age people jumping off from approximately the latitude of Britain and working their way around the icebound North Atlantic rim in some kind of watercraft. They would likely subsist by hunting sea mammals along the ice pack. The distance would not have been as great as it is today, because the Grand Banks off Newfoundland would have extended eastward at lower sea level.

But there are problems with this scenario as well. It entails making a long, rather fast journey skirting the ice pack, probably in a single summer season, since there would be no sheltered place to spend the winter. This would require a much more sophisticated boating technology than demanded by gradual North Pacific island-hopping over tens, hundreds, or even thousands, of years. But there is no evidence that the Solutrean people hunted sea mammals or had any other offshore maritime technology. In fact, no European evidence points to the existence of *any* advanced boats nearly that early. (The oldest known log canoe, from the Netherlands, is about 9000 years old, and it is far too small and its sides too low to embark safely on a long ocean voyage.) There is not even inferential documentation, such as evidence that people reached islands off northern Europe during that time period. Even in the much balmier Mediterranean, people first began colonizing the islands very close to shore 13,000 to 14,000 years ago, and did not reach more distant islands until 10,000 years ago or later. (Corsica, for example, which would be over the horizon, was not colonized until about 9000 years ago.)

In any case, large Solutrean or Clovis spear points are far from being the technology needed for hunting sea mammals on the edge of the ice pack. Furthermore, this "Solutrean connection" would imply an unbroken Clovis-style toolmaking tradition lasting at least 5000 years. But if people wielding such spear points lived in America all that time, why have none of their bones and teeth been found, or the spear points themselves for that matter? The allegedly pre-Clovis artifacts found at the eastern North American sites do not *look* like precursors of Clovis stone points. And if not, why not? The transatlantic theory has not gained many supporters.

Of all the alternative explanations for the peopling of the Americas, the least convincing of all is actually more of a philosophical, spiritual, even legalistic, viewpoint. This is the *a priori* assertion that Native Americans have "always" been here. As stated by one Washington State native leader involved in the Kennewick Man dispute, "From our oral histories, we know that our people have been part of this land from the beginning of time. We do not believe that our people migrated here from another continent."

This argument may be spiritually satisfying to native groups, and "politically correct" on the part of nonnatives who support their legal claims, but it defies everything we know of human evolution, which began with hominids in Africa and continued with much more recent regional evolution in Europe and Asia. Archaeologists in Africa and Eurasia have found myriad offshoots and precursors of modern humans, such as Australopithecines, *Homo erectus*, Neanderthals, and the like. But there has never been a trace of hominid ancestors in the Americas from which people here could have evolved. And if people *had* evolved independently in the New World, their genetic makeup would not have ended up essentially identical to that of Old World people, so that we are, and remain, all one species.

In the light of my visits to research sites and discussions with the scientists, very late Ice Age coastal migration seemed by far the simplest and most coherent way to account for the available evidence. And the picture for coastal migration kept getting better. However compelling it was, though, it was not yet proven. Even the scientists who were most firmly convinced about the scenario expressed themselves with caution. Biologist Rolf Mathewes, for example, would only say that the coastal migration route "could" have been used by humans early enough to account for Monte Verde. Archaeologist Jim Dixon stuck to the formulation that coastal migration was a "hypothesis." And Daryl Fedje would only call his 11,500- to 12,400-year-old pieces of sharp stone from the sea floor "possible" artifacts. A lot more solid evidence was needed before the coastal concept would be embraced wholeheartedly in the way that Clovis First had been.

Caution was a dominant theme at a symposium that Fedje convened at his Victoria, B.C. office in 1999 to review the years of research in the Queen Charlotte Islands and to discuss where the research effort should be directed next. Attending were nearly all the archaeologists, geologists, and biologists who had worked on the project for so many years. Representing the Haida were Tom Greene Jr. and Captain Gold.

A high point came when Heiner Josenhans placed on the conference table a dustpan holding the 12,200-year-old pine tree stump, still embedded in peaty soil, that had been dredged from 476 feet. Rolf Mathewes expressed relief to see an actual tree of that age. Until then, his reconstructions of the ancient "Lost World" environment had described pine forests beginning to replace the earlier tundra at around that time. But these had been based on studies of pollen from sediment cores, and pollen can blow long distances on the wind. Josenhans and Fedje also displayed samples of the clamshells as evidence of the abundant food resources available on the drowned coast back then.

As for the "possible" artifacts, though, the scientists were making no claims for them as evidence of anything. Additional archaeologists at Simon Fraser University had looked closely at these pieces of broken stone, and some certainly thought they could be artifacts. The trouble was that each archaeologist seemed to have a different "favorite" candidate for artifact status. There was no consensus.

Discussion turned to the question of where to look next. One possibility was simply to go back to Juan Perez Sound, if funding and ship time could be secured, and devote more effort to pounding the bottom targets with the grab. But there were alternatives. Captain Gold, who knew the geography of the Charlottes intimately, pointed out a number of sheltered nooks, harbors, and inlets on the western side of Moresby Island, where the sea bottom might well be worth investigating for ancient habitation. The sea level history and relationship to local glaciers there were somewhat different from the areas that had been targeted to date on the eastern side of the Charlottes. First would have to come the preliminary work, including sediment cores and swath bathymetry imaging of the seafloor.

Fedje was also interested in searching for ancient habitation sites on land along the mainland side of Hecate Strait, near the city of Prince Rupert. The Cordilleran ice sheet began receding from the coast by around 14,000 years ago, so it was possible that slightly later people had inhabited some shore sites there. It would have been relatively barren, recently deglaciated land, but there might have been shellfish, sea mammals, shore birds, and other food resources to sustain them.

The advantage of looking on the mainland shore was that the postglacial geological history there was quite different from the Charlottes. The massive weight of the ice sheet had depressed the land by many hundreds of feet. When the ice retreated from the coast, that land began to rebound, and the total postglacial rebound exceeded the simultaneous worldwide increase in sea level. This meant that any ancient habitation sites would now be hundreds of feet *above* sea level, that is, on dry land, terra firma. There would be no need for an underwater search.

The catch was that if such sites existed they would be far up steep slopes and well back from the sea in what is today dense coastal rain forest. It would be very difficult terrain to explore. Moreover, the alkaline chemistry of the rain forest would have destroyed any shells and bones, so only stone artifacts would likely have survived. (In two of the three main coastal research areas, Alaska and California, the most valuable finds to date had been human bones and a bone artifact, so this was no mere hypothetical problem.) It was not clear just where any concentrations of such artifacts on today's hillsides were likely to be. It would have to be in places where sea level had remained stable for extended periods late in the glaciation. Only spots that had been along beaches or streams meeting the sea, and had remained that way for decades or centuries, were likely to have an accumulation of telltale human artifacts or toolmaking debris. Such sites would be difficult to find. But in contrast to the underwater work from an expensive ship, to access these areas would only require good hiking boots and possibly a four-wheel-drive vehicle.

I came away from this seminar convinced that a complete picture of coastal migration would never be assembled until more and better evidence was found from the now-drowned coast itself, all along the Pacific. That is, from under the sea. I pondered what Don Morris had told me about where he would look for evidence of the earliest habitation on Santa Rosa Island. The best area to search would be on the island's sheltered side. But in California, just as in Alaska and British Columbia, sea level rose about 400 feet as the Ice Age waned, which pretty well ruled out using scuba divers for an underwater search. The water was too deep. I thought about what Fedje and Josenhans had been doing in the Queen Charlottes, banging away at likely ancient occupation sites with a set of jaws that could only bite in to a depth of about two feet.

I had also been in touch with geologist Tom Ager, who was pinpointing the late Ice Age refuges on the outermost coast of southeastern Alaska. It was possible, of course, that if there were people on those outer islands 13,000 to 15,000 years ago, a few of them might have left traces in locations, such as caves, that are now above sea level. But these places would be akin to the tip of an iceberg. Most of the likely shoreline habitation, hunting, fishing, and foraging sites used by ancient people would—as with the rest of the iceberg—now be deep underwater. Such spots are too deep to excavate by human divers, and extremely difficult to probe with the rather crude dredging jaws deployed from an incredibly costly research ship.

I gradually realized that if scientists were ever going to find solid archaeology to support coastal migration, a new and better technological approach was needed. For one thing, it had to be more refined than biting a chunk out of the sea bottom with dredging jaws. Dredging might succeed in bringing an artifact to the surface now and then, but it would never satisfy the high standards of terrestrial archaeology. These require showing the layered, and chronologically consistent, geological context in which the artifacts are found. Otherwise, there is no way to be sure an artifact did not intrude from above into a deeper (and older) geological stratum.

There was also the question of cost. If a project was spectacular, and able to generate enough public buzz, money might not be a consideration. Rediscovering the Titanic, for example, was sexy enough that the scientists involved could afford expensive research ships and highly sophisticated submersibles. But no private foundation or public agency was likely to fund such expensive toys for what could amount to decades of searching underwater for ancient artifacts.

Fedje and Josenhans had been able to secure a grand total of seven days of ship time in the late 1990s for their concerted underwater probing. They found one artifact. The *Vector* cost many thousands of dollars per day to operate, and other important projects and sciences were in direct competition for its facilities.

The situation farther north was even worse. As Tom Ager explained, the swath bathymetry has not yet been done for the relevant continental shelf of southeastern Alaska. There is no fleet of research vessels available to survey, map, and then systematically probe the sea bottom on the Alaska coast, nothing remotely similar to what the Canadian government did in the 1990s. This is an unusual case in which the Canadian scientific effort is at least a decade ahead of the American one. Ager believed that the U.S. Navy has done classified bottom surveys in some inlets, mainly to figure out where submarines might hide. But in the area of greatest interest to him, the marine geology effort has been zilch.

If the search for ancient coastal migration was to continue and achieve success, I reckoned, a cheaper but also more sophisticated technological approach had to be found: something that did not require a well-staffed and large research vessel to be stationed on site throughout the project; something that could continue working during bad weather and at night; a technology that could operate efficiently at depths of 400 to 500 feet; one that could probe much deeper into the seafloor, and in a more controlled and systematic way, than the crude grab jaws.

I thought of what Norm Easton and Tom Greene had told me, about how poor the visibility could become when digging stirred up the seafloor sediment. And about how, in both of the underwater digs in British Columbia, it was actually the airlift or water suction systems that did most of the useful work, bringing material to the surface for examination there.

I recalled Henk Don's little robot submersible, MURV, with its lights, cameras and robot arm. This kind of underwater technology was no longer exotic. The components did not have to be custom designed. Even the robotic manipulator arms could now be purchased essentially off the shelf. The same was true for the communications links that could operate such systems remotely. Surely, with satellite communications links, there was no longer a need for an expensive ship with a large crew to be stationed over the work site at all times, especially if no human lives were at risk. Besides, a ship would always be at the mercy of human schedules and the vagaries of weather.

What was needed instead was a way to conduct a systematic, controlled, extended, and well-documented underwater dig, just as had been accomplished by divers in two shallow areas on the B.C. coast. But it had to be done at great depth. This meant using remote-controlled technology, not humans. It had to be done without a costly ship required to hover on station throughout the project at a cost of $10,000 or more per day. And it could not be done by a roving robotic vehicle, like MURV, which was poorly equipped for digging and at the mercy of the currents.

What might work, I realized, was a heavy robotic digging frame that would sit for weeks at a time in one place on the sea bottom. It could be sized to excavate a much larger section of the seafloor than the grab jaws. I began to envisage a heavy rectangular frame—open in the middle—welded from steel girders six to ten feet long. This could be lowered to the bottom from a ship or barge, in fact from almost any vessel large enough to carry a heavy-duty crane. The vessel could wait for good weather to deploy the digging frame. This would make the system practical even for working out in places like Hecate Strait or the exposed outer coast of Alaska, where sea conditions are often terrible, but are occasionally benign.

The frame would be lowered to its target, probably a level spot along a drowned river delta, or just above and back from an Ice Age beach site. Along each side of the frame would be bright lights to illuminate the dig, and TV cameras to record the activity. Each side would also be equipped with telescoping sheet metal structures that could be extended downward into the sea bottom as the dig progressed. These would act like the caisson walls that proved to be so important in the B.C. underwater digs, preventing debris from infiltrating the excavation hole. Also attached to the frame would be one or more electric-powered pumps to run the suction system.

The most important equipment attached to the frame, though, would be a number of highly mobile and extendable robotic arms that could reach inward from the sides. One such arm (or arms—there might be one or more along each side), would hold a trowel or similar digging tool. With a TV camera watching, it could scrape up material from the seafloor. Another arm (or arms) would hold and manipulate the end of the suction hose. Powered by an electric pump, this would suck up the loosened material and shoot it into fine mesh bags located in a collection module suspended underwater a short distance above the frame. Each bag would hold the material collected from a unique square and depth level within the grid being excavated by the digging frame.

If scraping the soft material stirred up the bottom so badly that the system was blinded, work might have to stop until the current swept away the cloud of particles. But no humans would be down there, and no expensive ship stationed above. Moreover, the system could work twenty-four hours a day. So a slow and deliberate work pace would not be a handicap. Cameras could keep track of the exact location and depth of the digging within the frame. The system would gradually excavate the grid inside the frame, square by square and layer by layer, just as is done in terrestrial archaeology. If a large object, such as a rock, were found, a robot arm would have to remove it. Large objects of potential archaeological interest would be grabbed, placed into separate mesh bags or baskets, and held temporarily in the collection module. As the dig proceeded and the hole got deeper, the telescoping sheet metal sides would extend down into the bottom to protect the hole from intrusions of unwanted material.

The only thing floating on the surface would be a large buoylike object (or possibly a sealed, unmanned barge) firmly anchored above the dig site by cables leading out in three or four directions. Inside the buoy would be a diesel generator, fuel cells, or some other source of power that would provide electricity through a cable to the lights, pumps, robot arms, and cameras deployed far below. The buoy would also provide satellite links to a control room or laboratory on shore, from which technicians would monitor the entire system on a real-time basis in safety and comfort.

Periodically (probably only in daylight and good weather), a boat or barge would come out to service the buoy. The fuel tanks supplying the generator or fuel cells would be refilled. The collection module would be raised briefly to the surface so that the bags holding the precious excavated material could be retrieved and taken back to the home laboratory for proper screening and study. Empty bags would be put in their place, and the system would continue working. If it ran out of fuel before the service boat arrived, or if something else went wrong, the system could simply be shut down. When the dig was complete, the frame would be repositioned at the next excavation target.

With round-the-clock digging capability, no full-time service vessel required on station, and few people involved, this kind of system should be much cheaper than dredging from a ship. It would also be far more controlled and precise, and less subject to the vagaries of weather. None of the components required are beyond what is commercially available today. They would not even have to be miniaturized, or be made of particularly lightweight materials, which might add to the cost. And the system would permit a search for artifacts that lie much deeper in the bottom muck than grab jaws can reach. The initial investment would be substantial, but once it had been made, the digging frame should be able to function for many years with relatively modest maintenance and operating costs.

The potential payoff is nothing less than the chance to answer in a definitive way the most gripping question of American prehistory. The quest for evidence of coastal migration is just beginning. To assemble a thoroughly convincing picture of the earliest coastal habitation will require excavating a whole series of well-dated underwater sites as well as some on land. This will surely take decades of work, and there is no better time than now to begin.

CHAPTER EIGHTEEN

Ancient Odyssey

A SMALL GROUP of people in crude sealskin attire fanned out along the beach and mudflats of a sheltered bay. The children and elderly used sticks to dig clams and harvested seaweed in baskets. Some of the adults waded into the shallow water and stretched out nets to catch fish. A few hiked far off down the shore, gathered bundles of driftwood and carried them back to feed their cooking fires. One or two butchered a seal that had been harpooned from a raft just offshore. Others used strips of sealskin to lash together driftwood for a new raft.

After the day's toil, they gathered around their beach fire and lamented that the pickings had been slim. They had been harvesting clams on this bay and neighboring ones for many summers, and the shellfish beds were nearly exhausted. The driftwood, too, had been thoroughly culled and was becoming alarmingly scarce. They needed to lay in large stores of food and fuel to survive the long and harsh Ice Age winters. It was getting much harder to eke out a living in the area. And their population was growing.

Recently, though, one of their rafts, while out hunting sea mammals, had been swept by unexpected winds far to the north and east. It had fetched up on a shoreline where there were no old campfire hearths or other signs of previous human activity. Behind those distant beaches were thick windrows of seasoned, bleached-out driftwood, the accumulation of decades. When the crew dug in the intertidal zone, it was full of clams and other shellfish. On rocks exposed at low tide, mussels grew in thick clusters. They had found an entire stretch of virgin coastline, a bountiful new territory. They struggled for days to return home, paddling hard when they could make headway, and camping on shore when they could not, to await favorable winds and currents. The involuntary scouts made it back and reported what they had found.

These were nomadic people, without fixed settlements. So the decision came easily. If the clan stayed where they were, they would starve or end up fighting with a rival clan to the south, whose numbers were also growing and whose resources were stretched to the limit as well. The solution was to move on, beyond their familiar hunting, fishing, and gathering territory. They completed the new raft, repaired several older ones, and salvaged what they could of their tentlike sealskin shelters. Meanwhile, they watched the weather and waited for ideal conditions. The summer days were long and the temperature well above freezing. But still there was no sense taking unnecessary risks.

When the winds and cloud patterns indicated a spell of fair weather, the tiny flotilla set out, paddling steadily for hours at a time, or drifting with the wind whenever it propelled them northeastward. Progress was slow but steady. They stayed close to land, never losing sight of the shore. To the northwest were high mountains shrouded at times in clouds and encased in glacial ice. During the short northern nights they carried on. It was safer than trying to find a landing spot on an unknown coast. After several days at sea, a stretch of shoreline came into view that had the conspicuous landmarks they were watching for. The rafts approached warily and landed on the beach of a protected nook.

What the earlier crew had told them was all true. The tidepools and intertidal zone held a cornucopia of shellfish. Edible kelp and other seaweed grew in profusion. Driftwood lined the shore. It must have floated in from distant, unknown lands. Behind the beach only sparse and scrubby trees grew in a few sheltered places. Everywhere else was open tundra. But they were people of the sea who exploited few forest resources, and so they used the bones of large sea mammals instead of wood for many tools and weapons.

As the transplanted clan set up their shelters, gathered food, and made their first fires, they knew they had found a good new place to live. There were no rival clans to challenge them. At least not right away. So much food and fuel were close at hand that the first year or two in the region promised to be relatively easy living. As pioneers, they would have to explore the surrounding region and locate new sources for some essentials, such as toolmaking stone. They would have to seek out estuaries, where they could construct fish traps, and bird colonies, where they could harvest eggs. But overall, the climate and resources in their new home territory were not drastically different from the old one, except for the greater, untouched bounty. They were adept at living from the wealth of the sea and shoreline. It's what they had learned from their ancestors as far back as memory and legend could tell. As they went to sleep that first night, they enjoyed the security of knowing that they would not have to move onward for years to come.

This imagined reconstruction is the kind of picture that came to mind as I read the research reports, talked to scientists, and tried to envisage what a very early coastal migration around the North Pacific rim likely entailed. There were, I realized, lots of unknowns. Decades of additional research, including extensive undersea archaeology, will no doubt be required before we can assemble anything like a complete mosaic of the earliest peopling of the Americas. No one knows yet just who these first pioneer people were; what types of boats, rafts, or other watercraft they built; or precisely when their migration began. The earliest episode in this ancient American saga remains the elusive Holy Grail that it has been for a century.

But based on the coastal research so far, we can construct a credible scenario of ancient migration and population spread that fits the known and likely evidence.

For starters, the people involved would not have realized that they were colonizing a new continent and hemisphere. There is no need to assume any conscious intent or motive beyond the quest for survival and the need to move on to new and more resource-rich territories.

The migration would have begun on the northeastern coast of Asia, most likely around 15,000 years ago. The first migrants might have come from among the Jomon of Japan, who had a maritime culture by the last millennia of the Ice Age. Or perhaps they set out from the coast of Kamchatka.

Whoever these distant ancestors of the first Americans were, they were adapted already to a cold maritime environment. They were reasonably skilled in navigation, at least close to land. Their rafts (or perhaps skin boats, but probably not dugouts, since large trees were scarce on the Ice Age North Pacific coast) would be adequate for short voyages along the shore for fishing, hunting and other food gathering, possibly even for raiding or warfare.

Eventually, some of these people began to move farther north and east than their ancestors had ever ventured. By this time, the Ice Age was already waning. Both year-round glacial ice and seasonal shoreline pack ice were retreating or disappearing entirely from the North Pacific rim. The ameliorating climate opened new lands for habitation.

Theirs was still an extremely hostile environment in winter, when heavy pack ice lined the shore. They were forced to huddle in caves, or in dugout shelters roofed over with mammal skins, for much of the year, subsisting on dried and stored food. In summer, though, they could dig clams or gather mussels and roast or dry them for winter use. They gathered and dried vast quantities of seaweed. The coastal tundra offered the starchy edible roots of bistort and chocolate lilies. They could harvest fish in streams with fish traps or catch them from the beach in nets. Offshore, they hunted seals and sea lions, and possibly walruses, Steller's sea cows and narwhals as well. These mammals gave them meat and oil, skins for clothes, shelters and cordage, plus bone and ivory for harpoon heads and other implements. They also stocked up on driftwood for cooking and to heat their caves or other shelters.

At each new location, the farthest advanced group of migrants eventually felt a need to move on. This was mainly because the growing population put an excessive demand on resources. Shellfish beds became depleted. Driftwood had to be carried in from farther away each year. Generation after generation, one group or another leapfrogged ahead: first to the northeast along the coast of Siberia, then eastward along the southern coast of Beringia, and finally southward down the shores of what is now southeastern Alaska and British Columbia.

Along this entire route they encountered similar conditions and resources. Change was incremental and barely perceptible. No drastic new adaptations were required. There were shellfish and seaweed, sea mammals and tundra vegetation. Century upon century, the avant garde of migrants experienced the same incentives to move on that their parents and grandparents had faced and told them about.

Few of these hops, from one bay or offshore island stepping-stone to the next, took them far from shore. They could wait for good summer weather to venture forth in their rudimentary watercraft. Sometimes, though, they might be carried farther by the currents and winds than they intended. If so, their migration might move quite rapidly. In principle, people could have skirted the cold North Pacific rim in only a few years or decades to reach the unglaciated coast of today's Pacific Northwest. On the other hand, it might easily have taken them many centuries.

Most likely, they reached the glacial refuges of southeast Alaska and northern British Columbia by around 14,000 years ago. Much of the mainland shore was still glaciated, but this did not stop the people's advance. Flowing rivers of ice filled the deep channels between the Alaska islands, but these were retreating as much as a quarter mile per year. The formerly barren expanses of sterile bedrock and gravel were rapidly colonized by plants. Seeds arrived on the wind and in the guts of birds, or attached to their feet and feathers. Ocean currents spread shoreline plants rapidly as well.

Some migrants simply occupied the shores of these lush island refuges. Others kept on going, using the newly emergent and deglaciated islands as stepping-stones to move farther southward. More generations went by. Their descendants continued swiftly down the rugged outer coast of Washington and Oregon, where the shore was steep and there were few islands or sheltered harbors.

Until they reached these never-glaciated areas, it was impossible to move onto the mainland and into the interior of the continent. Ice blocked the way. But even when they were far enough south that there was no glacial ice, there was little incentive to penetrate the North American heartland. They were accustomed to coastal life and to exploiting coastal resources. The sheer abundance of shoreline food and fuel would remain an advantage however far south they progressed.

What prospects did the interior present? First, there was dense coastal forest. This led into steeply rising mountains with relatively few easy and safe passes through them. Beyond the mountains, from the Pacific Northwest to California and in much of Mexico as well, were open grasslands and in some places desert. The plants and animals in all these zones were entirely unfamiliar to coastal people. So were the ways to harvest and process those plants, or to hunt and trap those animals. New tools and toolmaking technologies would have to be developed to exploit these novel habitats.

It was simpler and safer to stick to their shoreline hunting and gathering than to brave the forests and mountains. Virgin shores continued to beckon them southward into ever more temperate climates, and then into the subtropics. Even in tropical climes, where inland conditions were unlike anything they had known, coastal life changed only gradually. It was something they could adjust to.

The narrow isthmus of Panama presented a major fork in the road for coastal migrants. Some venturesome group discovered the Caribbean Sea across that narrow neck of land and decided to follow those shores. In one direction, this took them along the northern coast of South America. By 13,000 years ago, some of them had learned to hunt mastodons. They left behind a spear point at Taima-Taima on the coast of Venezuela. In the other direction, they moved northward along the Gulf Coast of Central America, Mexico, and the southeastern United States. On the Atlantic shores, as on the Pacific, low Ice Age sea levels meant that the coastline they inhabited lies deep under water today.

Those who remained on the Pacific continued south and reached Chile by 12,500 years ago. It could easily have taken them 1000 to 1500 years to get there from the Northwest Coast. That would be forty to sixty generations. However, traveling by boat they might have done it much faster.

All this while, the small colonies of people left behind were growing in numbers, splitting off into new clans and larger population groups and diversifying genetically and linguistically. Yet they remained concentrated on the narrow but resource-rich coastal zone. And that zone grew larger as well. Along the mainland shore in the Northwest, the last of the ice sheets were gone by 12,000 to 13,000 years ago. Yet sea level was still very low. In fact, because the Cordilleran ice sheet had been pressing down the earth's crust, when it retreated the crust on the mainland shore rebounded, lowering relative sea level even farther and opening up additional coastal land to habitation.

As the climate warmed, new plant species became established where there had initially been only coastal steppe-tundra. By 12,500 years ago, pine forests began to replace the sedges, grasses, and scatterings of willows. And worldwide sea level began to rise, at first slowly but then quite rapidly. On the outermost coast, such as the Queen Charlotte Islands and the farthest offshore islands of Alaska, rapid settling of the earth's crust (the glacial forebulge) compounded the worldwide sea level rise.

By around 12,000 years ago, all along the northern Pacific coast the growing population and changing environment gave some people an incentive to experiment with new habitats and ways of exploiting them. The narrow coastal strip was becoming crowded. People there became more familiar with forest habitats. And so, the first small groups began extending their hunting and foraging territories far enough inland that they lost contact with the coast. Now adapted to interior ecosystems and lifeways, they moved into central Alaska and left the earliest Nenana artifacts there around 11,700 years ago.

Farther south they followed river valleys such as the Columbia, which led like highways deep into the interior of the continent. There they encountered the remaining Ice Age megafauna, such as the woolly mammoth and the bison. They found sources of fine toolmaking stone and learned to make large Clovis projectile points (something they never needed on the coast, where bone and antler harpoons were their main weapons) to hunt these huge beasts.

Meanwhile, one or more later surges, or waves, of migration brought new and biologically quite different people from northeastern Asia. Among them were people with the typical "Mongoloid" features of modern North Asians, who eventually made a large contribution to the genetic heritage of Native Americans. At some point, descendants of the original coastal people probably mixed, or fought, with later people coming south from Beringia either overland or via the coast.

Eleven thousand years ago, Arlington Woman, whose ancestors had been on the California coast for scores of generations, was out on a hunting or foraging expedition on the large Ice Age island of Santarosae. Perhaps she had an accident. Possibly she was killed by other humans. In any case, she died and her bones were preserved in the wet, boggy soil along with the bones of deer mice.

And on the Northwest Coast, hundreds of years later, another coastal dweller was probably hunting bears in a cave well inland and above the sea on Prince of Wales Island, Alaska. For some reason, he left a tool made of bone in the cave, to be discovered by a paleontologist more than 10,000 years later.

At almost the same time, someone was using a stone tool on the shores of a small river delta in the Queen Charlottes. He or she dropped that tool—a basalt knife—and it was left behind as the tide came up. Soon it was covered in sand, and within a decade the sea had risen so much that it was permanently under water. As sea level rose further, century after century, the stone blade lay just under the seafloor silt but in deeper and deeper water. Eventually, it was far from shore under Juan Perez Sound and was not seen again until huge steel jaws took a bite out of the sea bottom and hauled it onto the deck of a research ship.

When the Ice Age ended, people had been living on the Queen Charlottes, on the Alaska islands, and along the coastal strip right on down to South America for thousands of years. They had well-established trade routes for precious stone and probably other valuable materials. They were still nomadic, but all the most productive ecological niches in the coastal zone were being exploited, and a clan's hunting and gathering territories were well defined. Meanwhile, the environment continued to change rapidly. Spruce trees became established, mixing with, and gradually replacing, the predominant pines. Technology was changing, too, with the introduction of fine microblades and the tools and weapons that utilized them. Abundant large trees allowed dugout canoes to replace the former rafts.

Deep in the continental interior, unknown to the coastal dwellers, other people were learning to subsist with quite different methods and technologies. The Ice Age megafauna disappeared, due to overhunting or climate change. But across the Great Plains of today's United States, people pursued herds of bison. In the sparse northern woods and subarctic barren lands of Canada, they hunted caribou. Much farther afield, in the Near East, people were beginning to herd goats and sheep and to domesticate wheat and barley. This enabled them to establish the first fixed human settlements. It was not long before beans, corn, and squash were domesticated in Mexico and permanent settlements arose there, as well.

On the outermost North Pacific coast, the maritime gateway to the Americas, the original migration that had first brought people to those shores was now ancient history that was told in stories around a campfire. There were only dim and fading memories of heroic events and grim struggles for survival, but the accounts of that ancient era were gripping. Shrouded by the smoke from burning spruce logs, people passed on tales about how great expanses of ice, which were now only visible far up on the slopes of mountains, once came all the way down to the sea. Yarns were spun about a time before the coming of the first tree, when long-forgotten ancestors could walk from one island to the next across open grasslands. To some listeners gathered at the hearth, it sounded like pure fantasy.

Even in their own lifetimes, though, people on the outer coast were still experiencing the drastic changes caused by the ever-rising sea, which altered the location and character of the shoreline with each generation. The beaches, the estuaries, the shellfish beds, the shorebird colonies—all were shifting and migrating. As the sea encroached, nibbling away at the beaches and drowning fringe areas of low-lying forest, the coastal tribes were forced to seek out new and higher places to erect their shelters.

The elders also pointed out to sea and told about an earlier life on low-lying islands that were no longer visible. They told captivating tales—a rich mix of fact and fable—of the first places where their forebears had settled after migrating since the farthest reaches of living memory. Just as we find it so hard to imagine today, the children sat and listened, transfixed and scarcely believing that an entire coastal world that had existed for scores of generations could simply vanish beneath the waves.

INDEX

accelerator mass spectrometry (AMS), 7, 215, 251, 255
Acosta, Fray Jose de, 13
Adovasio, James, 151–52, 153, 158, 171, 190
Africa, hominids in, 271
Ager, Tom, 10, 23, 32, 274–75
Alaska Peninsula, 128
Aleutian Islands, 63, 74, 82, 128, 155
Aleut people, 74, 155, 156, 159
Algonkian people, 157
All Alone Stone, 227, 228, 230–34
Allred, Carlene, 29
Allred, Kevin, 6–7, 29, 30
American Museum of Natural History, 49
Amerind superfamily, 155, 156, 159, 160
animals, differences among, 19
Antevs, Ernst, 21
Anvil Cove, 109, 223–28, 235
Anzick site, Montana, 215
archaeology:
  careers in, 25, 150
  controversy and rivalry in, 147–54, 171, 179
  dating techniques in, 50–52; *see also* radiocarbon dating
  environmental indicators in, 240
  first excavation in, 15
  human remains as gold standard in, 215–16
  paleontology combined with, 35–36, 38
  repetition of findings in, 219
  standards of proof in, 148, 152–53, 214–15, 237, 265
  undersea, *see* undersea archaeology
Archipelago Management Board, 79
Arlington Canyon, California, 241–48, 251–54
Arlington Woman, 251–56
Arrow Creek, 117, 145
arrowheads, *see* spear points
artifacts:
  dating of, 50–52, 179, 215, 241, 263; *see also* radiocarbon dating
  geofacts vs., 152, 166
  positive identification of, 235–37, 238–40
  "possible," 201–3, 221–22, 231–32, 235–38, 271, 272
  see also specific sites
Asia:
  evolution in, 271
  human migration from, 21, 45–46, 156, 159, 264, 266
Athabaskan language, 155, 157–58
Atlantis, 14
atlatl, 50
Australia:
  colonization of, 122, 127, 169
  island-hopping around, 256
  linguistic diversity of, 161
Australopithecines, 271

Baichtal, Jim, 4–6, 8, 35, 173–74
Barb (nurse), 223, 228
Barker, Alex, 190, 191

Barrie, Vaughn, 102, 103, 112, 114, 115, 129, 131
barrows, 14–15
Barton, Benjamin Smith, 15
basalt, 202
bats, 28
beach terraces, 105
bears:
  denning spot for, 209, 262
  evolution of, 8–9
  fossils of, 7–8, 16, 23, 29, 30
  guns as protection from, 210–11
  migrations of, 6
Beauchamp, Richard, 85, 88, 89, 96
Bering, Vitus, 14
Beringia, 22, 74, 157, 166, 168
Bering Sea, water depth in, 20
Bering Strait, land bridge of, xiii, xvi, 11, 19–22, 25, 45, 51, 57, 62, 154, 165, 169, 187, 217, 265
Bessel, F. W., 17
Bettner, Rolf, 89
Bev (park warden), 87
biogeography, 19–20
Bird, Junius, 57, 150–51
Bismarck Archipelago, 122, 126
bison, 46–49, 259
Blackfoot language, 183, 184
blitzkrieg theory, 59
bones:
  dating of, 7–8, 16, 26, 30–32
  of humans, *see* human bones
  preservation of, 3
Bones, Boats and Bison (Dixon), 214–15, 218
Bonnichsen, Robson, 44, 45
Brace, C. Loring, 18, 264
Bradley, Bruce, 218, 269
Bringhurst, Robert, 258
British Columbia, map, x
Brooks Peninsula, 130, 132
Brown, Barnum, 49
Brown, Tucker, 96
Bryan, Alan, 147, 148, 163, 214
Bud (guide), 5
Buka Island, 123–26
Bumper Cave, 4–7
burins, 65
Burnaby Strait, 98–101

Cadorin, Grant, 99
camel, western, 259
Canadian Hydrographic Service, 198–99
Canadian Shield, 21
carbon 14 (radiocarbon), 53–57; *see also* radiocarbon dating
Carpenter, Janna, 35
Cave 49-PET-408, *see* On Your Knees Cave
cave paleontology, 3, 26, 29, 172, 205, 262
Channel Islands National Park, 241–44
Charlie Lake Cave, 68
Chatters, James, 39, 263
Cherokee people, 15
chimneys, 28, 29
Chisholm, Suzanne, 224
chlorine isotope, 186
Christensen, Tina, 81, 90
Chukchi Sea, 21
Chumash people, 245, 253
Clague, John, 74
Clovis First, 57–60, 166, 271
Clovis Mafia, 148–50, 153, 189–91
  critics of, 59, 68
  evidence sought for, 57, 61, 63
  and migration theories, 57, 58, 59, 61–63, 67, 130, 156, 181–82, 255–56, 261
  negative findings of, 62–63, 66–68, 75–76, 215, 265

other sites compared to, 57–58, 63–64, 241, 255–56
pre-Clovis sites, 62–63, 66, 147–54, 163, 169, 171, 187–91, 214–17, 255, 269, 270
repetition of findings, 219
Clovis points, 149
dating of, 49–51, 57; *see also* Clovis First
discovery of, 49–51
fluting of, 50, 63, 167–68
and migration theories, 57, 216, 269
in Paleo-indian tradition, 66
coastal hopping, 128
coastal migration, 80, 83, 169–71, 183, 258–66
Arlington Woman and, 255–57
from Asia, 21, 45–46, 156, 159, 264, 266
Clovis First and, 57, 58, 59, 61–63, 67, 130, 156, 181–82, 255–56, 261
coastal entry model of, 163
food resources in, 131, 257, 259–60
glacial ice and, 128–29, 130
ice-free corridor and, 22, 57, 62–63, 66, 68, 71–72, 130, 181–87
imagined reconstruction of, 259-62, 279–88
Kamchatka and, 264
and lack of human remains, 62–63, 66–68, 215, 265
languages and, 163
Monte Verde and, 169–71, 181, 191, 260
Quebrada sites and, 261–62
resistance to theory of, 25, 121, 130–31, 148, 149, 179
sea level and, 72–74, 158, 165–66, 168, 261
three migrations theory, 154–56, 263
transoceanic vs., 218
undersea evidence for, 129–31, 135–38, 173, 214, 215, 219, 274, 275
watercraft technology and, 121–23, 127–28, 169, 256
wide endorsement of, 214, 218, 264–66
cobble chopper, 236
colonization, minimum requirements for, 267–68
Conway, Kim, 98–101, 102–3, 105, 111, 116, 134–35, 144, 195
Cook, Captain James, 14
Cook Bank, 103, 131, 132, 197
coprolites, 188
Cordilleran ice sheet, 9, 21, 23, 24, 107, 181, 182, 272
Corsica, colonization of, 270
cosmogenic chlorine 36 dating, 186–87
Coupland, Gary, 190
Creation myths, 12, 13, 14–15
Cro-Magnon fossils, 265

Darwin, Charles, 17, 18, 25
Dawson, George M., 19–20
decay emissions, 54
deep space, 17
deep time, 17
Demmert, Rosanne, 177
Denali tradition, 65
dendrochronology, 51
dendrograms, 159
De Soto, Ferdinand, 14
diatom analysis, 105
digital elevation models, 193–94

INDEX

Dillehay, Tom, 150–51, 168–69, 171, 187–91, 214, 237, 266
Dincauze, Dena, 190, 191
Dixon, Jim, 24–25, 37, 165–80, 204–20, 265, 266, 271
 and blood residues, 124
 books by, 170–72, 173, 214–15, 217–18
 early years of, 165–66
 and Monte Verde, 169–72, 189
 at On Your Knees Cave, 35, 172, 173–80, 204–14
 and stone tools, 64, 124, 166–68
DNA testing, 157
Dogfish Bank, 133, 197
Don, Henk, 103, 116, 138, 194, 275
Duk-Rodkin, Alejandra, 182
Dyuktai tradition, 65

Easton, Norman, 135–38, 194, 195, 275
Echo Harbor, 134, 195
El Capitan Cave, 27–30, 32
Enlightenment, 14
Erlandson, Jon, 253, 255
erratics, 183–87
Eskimo-Aleut group, 74, 155, 156
Eskimos, 154, 157–58, 159
Europe:
 evolution in, 271
 genetic markers of, 160
 geological dating in, 49
 Neanderthals of, 46, 49, 215–16, 265, 271
 Solutrean tool tradition in, 218, 269, 270
 transatlantic crossings from, 269–71
"Eve," 159
evolution, 8–9, 157–58, 161, 263, 265, 271
extinctions, 58, 69, 121, 267–68

Fedje, Daryl, 77–84, 95
 and artifacts, 77, 81–83, 101, 110, 135, 144–45, 173, 192, 196, 202–3, 208, 231–32, 235–38, 271, 272
 and core sampling, 109, 117, 131, 197–98
 and dredging, 143–45, 192, 195–97, 230–32, 274
 at Echo Harbor, 134
 and Gwaii Haanas, 78–81, 110, 119, 131, 144
 and *John P. Tully*, 131, 197–98
 at Matheson Inlet, 82, 109, 145
 on migration theory, 83, 183
 papers published by, 80
 and Parks Canada, 78, 79, 80, 103, 114, 200
 at Queen Charlotte sites, 77, 78, 82, 103, 106–7, 131, 173, 175, 178, 192, 202, 212, 262, 271–73, 274
 at Richardson Island, 77, 134–35, 144, 145, 226
 on sea level history, 84, 105–7, 116–17, 202
 and totem pole project, 88–90, 92, 93
 and undersea exploration, 84, 116, 134, 138, 194, 195, 198, 199–201, 219, 262, 275
 and *Vector*, 77–78, 83, 99, 109, 116, 134, 143, 192, 194, 195, 197, 199, 221–23, 275
Fell's Cave, Patagonia, 58, 150
Fifield, Terry, 41–42, 174, 176–78, 211
Figgins, Jesse, 46–49
First Americans, 45–60
 in Clovis, New Mexico, 49–51, 57–60; *see also* Clovis First

and coastal migration, 73–74, 163, 171, 173, 181; *see also* coastal migration
dating techniques for, 49, 51–52, 53–57; *see also* radiocarbon dating
in Folsom, New Mexico, 48–51
Indians, 45–46, 154–56
in Kansas, 48, 49
long chronology of, 154–58, 263, 265, 266
at Meadowcroft site, 151–52, 158, 171, 217
older evidence for, 213–14
short chronology of, 216–17, 265
in Texas, 46–48, 49
see also South America
Fladmark, Knut, 68–74, 81–82, 135, 170, 181, 182, 266
"Fledermaus" computer system, 233
Flood-Tide Woman, 13
fluting, 50, 167–68
Folsom points, 48–51, 57, 62–63, 66, 149, 168, 219, 256
foothills erratics, 183–87
Forest Service, U.S., 4, 6, 7, 28, 29, 38, 41, 176, 177
foxes, arctic, 32–33, 262
Frydecky, Ivan, 102, 112–16
future, vision of, 276–78

Garrett, Ken, 224–25, 227, 228
genetic clock, 158
genetic diversity, 268
genetic drift, 263
genetic markers, 157–58
mtDNA, 158–60, 263, 269
genetic mutation, 161
geochronology, 214
Geological Survey of Canada (GSC), 19, 84, 103–5, 182, 183
Geological Survey of U.S. (USGS), 10
geology, 16–22, 49, 103, 197
geomorphology, 233
Ginsberg (elephant), 43–45, 189
glacial forebulge, 107
glacial man, 49
glacial moraines, 18, 20
glacial refugia, 30, 32, 76, 78, 132
glaciologists, 20
Gladstone, Ernie, 91, 93–97
glyptodonts, 147
Gold, Captain, 12, 91, 271, 272
Goose Bank, 131, 132, 197
GPS satellite system, 199
Grady, Fred, 30, 31, 33, 38, 174, 262
Graham Island, 70, 74, 78
Grand Banks, 269
Gray, Bob, 27–28
Greenberg, Joseph, 154–55, 156
Greene, Tom Jr., 12, 271
and diving, 195, 221–22, 228, 275
and human remains, 93, 96–97
and totem pole project, 88–91, 93, 96–97
trees felled by, 88–89
and undersea artifacts, 221–22, 235
Griffin, James, 168, 248
Griffiths, Alan, 262
ground truthing, 112
Gruhn, Ruth, 147–49, 162–63, 214
Guidon, Niéde, 152–54, 171
Gulf of Alaska, 128
Gwaii Haanas, 78–81, 97, 103, 110–11, 119, 131, 144

Haida Gwaii, 13, 68
Haida Indians, 11, 68, 70
archaeologists, 81, 82, 84, 90–92, 200, 210, 271
divers, 195–96, 221–22, 228

heritage sites of, 80
and human remains, 39, 93, 96–97, 176, 178
language of, 155, 158, 162
mythology of, 12, 13, 75, 258–59
origins of, 12
and park reserve, 79
totem poles of, 12, 85–97, 226
villages of, 78–79, 85
half-life, 54
Hamilton, Tom, 24
Harding, Jennifer, 230, 232–33
Haury, Emil, 244, 248
Hawaii, 120–21
Haynes, C. Vance, 67–68, 148, 190–91
Heaton, Julie, 2, 30, 38
Heaton, Tim, 1–10, 26–27, 29–39, 78, 164
and animal bones, 6, 7–9, 10, 16, 23, 29–32, 37, 38, 174, 262
and Bumper Cave, 4–7
and Forest Service, 4, 41, 176
and human remains, 38–39, 41–42, 175, 176, 179, 202, 211–12
and migration theory, 23–25
and National Geographic, 38, 222
and On Your Knees Cave, 1–4, 7–10, 25, 29, 30–39, 172–76, 179, 180, 205, 211
and spear points, 37
Hebda, Richard, 130
Hecate Strait, 68–69, 107, 108, 110, 131, 272
hematite, 209
Henderson, Jamie, 229
Heusser, Calvin, 71
Heyerdahl, Thor, 266, 268
Holocene epoch, 37, 74, 246, 253, 261
Homo erectus, 216, 271
Honshu Island, 127
horses, 58, 147, 259
Hrdlicka, Ales, 18–19, 46, 49
Hulten, Eric, 21–22
human bones:
Arlington Woman, 251–56
dating of, 212, 252
as gold standard, 215–16
Kennewick Man, 38–42, 160, 176, 178, 263, 264, 270–71
in On Your Knees Cave, 38–39, 175–79, 211–12, 252
of Sgan Gwaii, 92–93, 96–97
Hutton, James, 16–17

Ice Age:
and Bering land bridge, 57; *see also* Bering Strait
coastal habitation in, 73–74, 84, 238
environmental changes in, 255
in Europe, 265
and human migration, 22–23, 25, 121, 170, 264–65, 269–71
hunters in, xv, 66, 189, 214
last glaciation of, xiii-xiv, 49
mammals in, 8, 51, 58, 245, 259
people in, 10, 22, 46, 170, 274
radiocarbon dating for, 56–57, 178
and sea level, xiv, 9–10, 21, 70–74, 84, 100, 104–10, 116–17, 143, 158, 168, 234, 261, 272
timing of glaciations in, 27
tools of, 63–66
ice-free corridor, 128
and Clovis people, 216
food resources in, 72, 131, 182–83, 238, 259–60

and foothills erratics, 184–87
and human migration, 22, 57, 62–63, 66, 68, 71–72, 130, 181–87
and sea level, 71–72
Indians:
burial mounds of, 14–15, 18
genetic studies of, 157–58, 159
and human remains, 39–41, 92–93, 96–97, 175–78, 263, 264, 270–71
languages of, 15, 16, 155, 157–58, 162
migration theories of, 45–46, 154–56, 159
origins of, 12–13, 14–15, 40, 75, 270–71
shell middens of, xi-xii, 64, 70
teeth of, 19, 156
see also specific tribes
Inside Passage, xii, 103
intervisibility, 125–26
Inuit people, 155
Irwin, Geoffrey, 124–27
isostatic rebound, 107

Jackson, Lionel, 183–87
Jasper National Park, 184
Jefferson, Thomas, 14–15, 16
John P. Tully (research ship), 103, 131, 197–98
Johnson, John, 251
Johnston, W. A., 21, 22
Jomon people of Japan, 18, 264
Josenhans, Heiner, 175, 183, 221–22, 223
and artifacts, 77, 201–2, 231, 236, 238, 272, 275
and core sampling, 109, 129, 131, 197–98
and dredging, 143–44, 197, 229–30, 235, 238, 274
and marine surveying technology, 84, 98–99, 100–105, 111–16, 140–44, 192–94, 197–200, 219, 230–34
papers published by, 80
and sea level history, 106–10, 234
and undersea mapping, 84, 98, 102, 104–5, 111, 198, 233–34
Juan Perez Sound, 198, 199, 200, 222, 226, 232–33, 272

Kamchatka peninsula, 264
Kansas, Logan County, 48, 49
karst topography, 173–74
Keddie, Grant, 64
Keewatin dome, 21
Kenai Peninsula, 128
Kennewick Man, 38–42, 160, 176, 178, 263, 264, 270–71
Kilu Cave, 123–24
Klein, David, 8
Kodiak Island, 72–73, 128–29
Kon Tiki, 266, 268
Korff, Serge, 53, 54
Kozushima Island, 127
Krieger, Alex, 248
Kunghit Island, 93
Kushtaka Cave, 37

Labrador dome, 21
lag surface, 135
Lambert, Phil, 224, 225, 227–28, 234
languages:
and coastal migration, 163
evolution of, 161, 265
and genetic markers, 157–58
and long chronology, 154–56, 161–63
and three migrations theory, 154–56
Lapita pottery, 123
Laskeek Bank, 104–7, 110, 133, 197

INDEX

Laughlin, William, 74
Laurentide ice sheet, 21, 23, 181, 182, 185, 187
LaVigne, Roger, 140
Leakey, Richard, 44
Lee, Craig, 172, 205–6, 207
Libby, Willard Frank, 54–57
limestone caves, 172, 174, 215, 262
Lineage X gene string, 160
lithosphere, 107
Little, Edward, 186
long chronology:
  controversy on, 147–54, 215–17, 263, 265, 266
  genetic markers, 157–60
  linguistics studies, 154–56, 161–63
  short chronology vs., 216–17
  three migrations theory, 154–56
lost world, 130, 272
Lou Island, 126
Love, David, 37
Loy, Thomas, 124
Luternauer, John, 103
Lyell, Charles, 16–18
Lyell Island, 110–11
Lynch, Thomas, 148, 149

McArthur, Norma, 267
MacDonald, Glen, 75
McJunkin, George, 48
Mackie, Quentin, 228, 238–40
McPhee, John, 17
McSporran, Joanne, 77–83, 99, 103, 106–7, 112–13, 116, 135, 143–45, 199–202
Magne, Marty, 89, 92–94
mammoth bones, 44–45, 51, 248
mammoths, 63, 214, 246, 254–55, 259
Mandryk, Carole, 182–83, 191
Manger, Heidi, 206, 207–8
Mann, Daniel, 128–29
Manus Island, 126
marine reservoir effect, 212–13
Martin, Paul, 58–59, 73
mastodons, 147, 148, 151, 188, 214, 259
Mather, Cotton, 14
Matheson Inlet, 82, 109, 140, 145, 200
Mathewes, Rolf, 74–75, 80, 129–30, 183, 226, 271, 272
Matson, R. G., 190
Meadowcroft, Pennsylvania, 151–52, 158, 171, 217
Melanesia, 122
Meltzer, David, 130–31, 149, 190, 191
microblades, 64–66, 204, 213
Middle Bank, 131, 132, 197
migration theory, *see* coastal migration
Minthorn, Armand, 40
Mitchell, Donald, 136
Mongolia, migration from, 159
monkeys, Old World, 19
montane glaciers, 184
Monte Verde, Chile, 150–51, 169–72, 181, 187–91, 214, 216–17, 237, 255, 260, 266, 271
Moore, Charles, 195–96
Moresby Island, 82, 86, 104, 109, 117, 134, 272
Morlan, Richard, 44, 189
Morris, Don, 241–48, 250–57, 273
Morton, Samuel, 18
Mount Edziza, 205
Mount Vesuvius, 56
mowing the lawn (seafloor), 111, 112
Mrzlack, Heather, 206–7
mtDNA (mitochondrial DNA), 158–60, 263, 269
multibeam sonar, 198

MURV submersible, 138–43, 194, 275–76
museums, need for, 253

Na-Dene superfamily, 155, 156, 158, 159
NAGPRA (Native American Graves Protection and Repatriation Act), 40, 176
National Geographic, 222–25, 228
National Geographic Society, 38, 151, 175, 222
National Science Foundation, 37, 175, 179–80
Native Americans, *see* Indians
Neanderthals, 46, 49, 215–16, 265, 271
Nenana, 167, 168, 169, 171
Neves, Walter, 263
New Britain, 123
New Guinea, 123, 161
New Ireland, 123, 124–25
Neynaber, Joe, 28
Nichols, Johanna, 161
Ninstints, 85
Northwest Passage, 14
Nozaki, Kei, 1, 3, 33, 34
nunataks, 21

obsidian, 37, 123, 204, 205, 208, 209
ocher, 208
Okinawa, 127
Okotoks (Big Rock), 183–84
Olsen, Patrick, 209, 210–11
On Your Knees Cave, 1–4, 25, 29, 30–39, 132, 172–80, 204–14
  animal bones in, 7–10, 16, 30, 31–33, 37, 38
  Bear Passage in, 3, 173
  human bones in, 38–39, 175–79, 211–12, 252
  as landmark site, 179–80
  official designation of, 4, 6
  Seal Passage in, 1–2, 31
Orr, Phil, 245–51, 254–55
otter:
  bones of, 82
  ceremonial burial of, 29
overkill theory, 59
Owen, Bessie, 245

Paleoarctic tradition, 64, 66
Paleo-indian tradition, 66
paleontology:
  archaeology combined with, 35–36, 38
  cave, 3, 26, 29, 172, 205, 262
Parfit, Mike, 224
Parks Canada, 78, 79, 80, 85, 103, 114, 195–96, 200
Parrish, Eric, 206
peat analysis, 232
Pedra Furada, Brazil, 152–54, 171, 216, 265, 266
peripheral bulge, 107
Peru, archaeology in, 261–62
Peteet, Dorothy, 128–29
Phillips, Steve, 68–69
photosynthesis, 54
plate tectonics, 20
Plato, 14
Pleistocene epoch, 45, 73, 121–22, 127–28, 132, 147, 151, 185, 246, 249, 256, 261; *see also* Ice Age
pollen samples, 26, 71, 74, 84, 103, 232, 272
Polynesians, 58, 120–27, 266–67, 268–69
Pompeii, 56
Powrivco Bay, 113–14
Pribilof Islands, 19, 20, 165–66
Prince of Wales Island, 1–11, 12, 26,

29, 31–32, 72–73, 164, 173–74, 176, 178, 205, 215, 262
pygmy mammoth bones, 246, 248–50, 254–55, 255

quartz crystals, 208, 209
quartzite, 183–87
Quebrada Jaguay, Peru, 261–62
Quebrada Tacahuay, Peru, 261–62
Queen Charlotte City, 69, 70
Queen Charlotte Islands, 68–72, 77, 78, 103, 175, 192, 274
  animal species of, 69, 262
  artifacts in, 70, 82, 173, 192, 202, 212
  conference about, 271–73
  as glacial refuge, 72, 74–75
  Haida people of, 12, 68, 70, 75, 178
  and sea level, 70–72, 106–8, 131, 173
Queen Charlottes Museum, 92
Quest for the Origins of the First Americans (Dixon), 170–72, 173, 217–18

radiocarbon dating, 7–8, 51, 53–57, 65, 83, 178, 212, 215, 218, 232, 249
Rainey, Froelich, 52–53, 55–56, 61–64
Raven myth, 75
reconstruction, imagined, 259–62, 279–88
Reitmeyer, Dylan, 204, 206
Reitmeyer, Lynn, 204
Rice, Marlene, 68–69
Richardson Island, 77, 134–35, 144, 145, 226
Ringer, Jim, 195, 196
robot submersibles, 138–43, 194, 275–76
Rocky Mountains, foothills erratics of, 183–87
Rogers, Richard A., 161-62
ROV (Remotely Operated Vehicle), 138–43
Rowsell, Barb, 225–26, 228
Rowsell, Keith "Flash," 225–28

Salish tribes, xii
Santa Barbara, California, 241–45
Santa Rosa Island, 242, 244–50, 254, 273–74
sea caves, 174
sea floor:
  dredging of, 83–84, 112, 143–45, 192, 195–97, 199–201, 202, 229–32, 235, 238, 274
  mapping of, 84, 98–99, 102, 109, 111–15, 193, 194, 197–99
  sediment cores taken from, 84, 98–101, 103–5, 109–10, 112, 114–17, 129, 131, 197–98, 219, 229, 272
Sealaska Corporation, 177, 180
sea level:
  eustatic, 107–8
  and human migration, 72–74, 158, 165–66, 168, 261
  Ice Age, xiv, 9–10, 21, 70–74, 84, 100, 104–10, 116–17, 143, 158, 168, 234, 261, 272
seals, 19, 31, 32, 260, 262
seamounts, xiv
sediment transport, 233
Sgan Gwaii, 12–13, 85–97
shellfish, 259–60, 272
shell middens, xi-xiii, 64, 70
Shields, Gerald, 8
Shimamura, Kazuharu, 186
sidescan sonar, 111, 198
Sirles, Wade, 34–35
Skidegate Inlet, 69

Skoglund's Landing, 70
Smith, Pete, 172–73
Smithsonian Institution, 30, 38, 190
Sneferu, pharoah, 56
Solomon Islands, 123, 169
solution caves, 174
solution tubes, 2
Solutrean tool tradition, 218, 269, 270
South America:
Fell's Cave, 58, 150
migration to, 59, 163, 191, 261, 263, 267–68
Monte Verde, 150–51, 169–72, 181, 187–91, 214, 216–17, 237, 255, 260, 266, 271
Pedra Furada, 152–54, 171, 216, 265, 266
Peru, 261–62
pre-Clovis evidence in, 148–51, 152-54, 215
Taima-Taima, 147–48, 214, 217, 260
Southon, John, 212
spear points, 37, 38
Clovis, 49–51, 57, 63, 66, 149, 167–68, 216, 269
Folsom, 48–51, 57, 62–63, 66, 149, 168, 219, 256
in Kansas, 48, 49
in Texas, 47–48
spokeshave, 208
Stafford, Tom, 178, 252–53
Stalker, Archie, 66, 184
Stanford, Dennis, 44, 190, 218, 269
Steele, D. Gentry, 264
Stellar's sea cow, 260
Stevens, Millie, 176
Stone Age, Neanderthals in, 215–16
Stone House, xi
stone tools, 5, 44, 49, 57–58, 64, 110, 124, 147, 151, 166–68, 188, 192, 238–40, 261–62; *see also* spear points
Story as Sharp as a Knife, A (Bringhurst), 258–59
subbottom profiler, 105, 111–12, 114–16
Sumpter, Ian, 80, 90–91, 200, 201
swath bathymetry, 198–99, 200, 230, 275
Szathmary, Emöke, 156–58

Taima-Taima, Venezuela, 147–48, 214, 217, 260
Taylor, Jim, 89
teeth:
of Indians, 19, 156
pre-Columbian, 155–56
Texas, Lone Wolf Creek bones, 46–49
Thomas, David Hurst, 178
three migrations theory, 154–56, 263
Timor, 122
Tlingit Indians, 5, 11, 210
and human remains, 39, 175–78
language of, 155, 158, 162
Tongass Cave Project, 29
Tongass Forest District, 4, 29
totem pole preservation, 85–97, 226
trade networks, 208, 213
transoceanic migration, 160, 218, 266, 269–71
tree ring dating, 51
Turner, Christy II, 155–56

Umatilla people, 40, 176
Unalaska, 20
undersea archaeology, 220, 228–40, 274–78
with *Anvil Cove*, 109, 223–24
core sampling, 84, 98–101, 103–5,

109–10, 112, 114–17, 129, 131, 197–98, 219, 229, 272
dredging, 83–84, 112, 143–45, 192, 195–97, 199–201, 202, 229–32, 235, 238, 274
on *John P. Tully*, 103, 131, 197–98
and scuba divers, 103, 116, 134–38, 194–97
technology of, 98–99, 100–105, 111–16, 138–46, 192–94, 197–200, 219, 230–34
and tides, 82–83
on *Vector*, *see Vector*
undersea fiber-optic cable, 192–93
undersea mapping, 84, 98–99, 102, 104–5, 109, 111–15, 193, 194, 197–99, 233–34
Ushki Lake, 264
Ussher, James, 15, 16

Vaara, Yarrow, 209–10
Vancouver Island, 104
varve dating, 51–52
Vaughn, Nelson, 46
Vector (research ship), 77–78, 80, 134, 194–95, 197, 221–40
administration of, 229
core sampling with, 98–101, 129, 229
cost of, 100, 109, 275
dredging with, 83, 192, 199, 229–30
laboratory instruments of, 98, 111–12, 116, 230–34
and MURV, 138–43, 194
undersea mapping on, 98–99, 111, 113
Vesuvius, eruption of, 56
vibracore, 100, 197

Waddell, Pete, 103, 116, 134–35, 140–41, 195
Walker, Phillip, 252
Warner, Barry, 74
West, Frederick H., 66–68
White, Rob, 113, 134, 141
Wickler, Stephen, 123–24
Wilson, Barb, 89, 90, 93–95, 97
Wilson, Bert, 195
Wisconsin glaciation, late, 22, 187
Wrangell Island, 58
Würm, 22

Yeltatzie, Jordan, 91–92, 93, 200, 201
Young, Sean, 92
Yuit people, 155

Zeuner, Frederick, 53